Springer Theses

Recognizing Outstanding Ph.D. Research

For further volumes:
http://www.springer.com/series/8790

Aims and Scope

The series "Springer Theses" brings together a selection of the very best Ph.D. theses from around the world and across the physical sciences. Nominated and endorsed by two recognized specialists, each published volume has been selected for its scientific excellence and the high impact of its contents for the pertinent field of research. For greater accessibility to non-specialists, the published versions include an extended introduction, as well as a foreword by the student's supervisor explaining the special relevance of the work for the field. As a whole, the series will provide a valuable resource both for newcomers to the research fields described, and for other scientists seeking detailed background information on special questions. Finally, it provides an accredited documentation of the valuable contributions made by today's younger generation of scientists.

Theses are accepted into the series by invited nomination only and must fulfill all of the following criteria

- They must be written in good English.
- The topic should fall within the confines of Chemistry, Physics, Earth Sciences, Engineering and related interdisciplinary fields such as Materials, Nanoscience, Chemical Engineering, Complex Systems and Biophysics.
- The work reported in the thesis must represent a significant scientific advance.
- If the thesis includes previously published material, permission to reproduce this must be gained from the respective copyright holder.
- They must have been examined and passed during the 12 months prior to nomination.
- Each thesis should include a foreword by the supervisor outlining the significance of its content.
- The theses should have a clearly defined structure including an introduction accessible to scientists not expert in that particular field.

Andrew James Logsdail

Computational Characterisation of Gold Nanocluster Structures

Doctoral Thesis accepted by
the University of Birmingham, UK

Springer

Author
Dr. Andrew James Logsdail
Department of Chemistry
University College London
London
UK

Supervisor
Prof. Roy L. Johnston
School of Chemistry
University of Birmingham
Edgbaston, Birmingham
UK

ISSN 2190-5053 ISSN 2190-5061 (electronic)
ISBN 978-3-319-34865-0 ISBN 978-3-319-01493-7 (eBook)
DOI 10.1007/978-3-319-01493-7
Springer Cham Heidelberg New York Dordrecht London

Printed on acid-free paper

Springer is part of Springer Science+Business Media (www.springer.com)

Parts of this thesis have been published in the following journal articles:

"*Theoretical and Experimental Studies of the Optical Properties of Conjoined Gold-Palladium Nanospheres*". Andrew J. Logsdail; Nikki J. Cookson; Sarah L. Horswell; Z. W. Wang; Z. Y. Li; and Roy L. Johnston. *J. Phys. Chem. C*, **2010**, 114 (49), pp 21247–21251
* Reproduced with permission

"*Interaction of Au_{16} Nanocluster with Defects in Supporting Graphite: A Density-Functional Study*". Andrew J. Logsdail; and Jaakko Akola. *J. Phys. Chem. C*, **2011**, 115 (31), pp 15240–15250
* Reproduced with permission

"*Development and Optimisation of a Novel Genetic Algorithm for Identifying Nanoclusters from Scanning Transmission Electron Microscopy Images*". Andrew J. Logsdail; Z. Y. Li; and Roy L. Johnston. *J. Comp. Chem.*, **2012**, 33 (4), pp 391–400
* Reproduced with permission

"*Interdependence of Structure and Chemical Order in High Symmetry $(PdAu)_N$* Nanoclusters". Andrew J. Logsdail; and Roy L. Johnston. *RSC Adv.*, **2012**, 2 (13), pp 5863–5869
* Reproduced with permission

"*Predicting the Optical Properties of Core-Shell and Janus Segregated Au-M Nanoparticles (M = Ag, Pd)*". Andrew J. Logsdail; and Roy L. Johnston. *J. Phys. Chem. C*, **2012**, 116 (44), pp 23616–23628
* Reproduced with permission

"*Faceting Preferences for Au_N and Pd_N Nanoclusters with High-Symmetry Motifs*". Andrew J. Logsdail; Z. Y. Li; and Roy L. Johnston. *Phys. Chem. Chem. Phys.*, **2013**, accepted (doi:10.1039/C3CP50978H)
* Reproduced with permission

I have also contributed to the following publications:

"*Dopant-induced 2D-3D transition in small Au-containing clusters: DFT-global optimization of 8-atom Au-Ag nanoalloys*". Sven Heiles; Andrew J. Logsdail; Rolf Schäfer; and Roy L. Johnston. *Nanoscale*, **2012**, 4 (4), pp 1109–111

"*Overgrowth of Rhodium on Gold Nanorods*". Ruth L. Chantry; W. Siriwatch-arapiboon; Sarah L. Hosrwell; Andrew J. Logsdail; Roy L. Johnston; and Z. Y. Li. *J. Phys. Chem. C*, **2012**, 116 (18), pp 10312–10317

Supervisor's Foreword

Gold nanoparticles and gold-containing nanoalloys have been of great scientific interest since the discovery of their novel catalytic activity in the 1990s. Thanks to advances in size-selected synthesis and electron microscopy techniques, characterisation of the shape and composition of such nanoparticles has become possible in recent years. However, gaps remain in our understanding of the structural and electronic factors which could lead to the generation of better noble metal catalysts.

Computational chemistry is a powerful tool in contemporary nanoscience: complementing experimental observations, allowing the prediction of potential new catalysts for further study and helping to guide future experiments. In this Ph.D. thesis, Andrew Logsdail presents the results of computational studies that aid the understanding of structural trends in gold nanoparticles and compositional and chemical ordering preferences in bimetallic nanoalloys formed with other noble metals, such as palladium and platinum. Highlights include: calculations of the structural preferences and optical response of gold nanoparticles and gold-containing nanoalloys; the design and implementation of novel numerical algorithms for the structural characterisation of gold nanoparticles from electron microscopy images; and electronic structure calculations investigating the interaction of gold nanoparticles with graphene and graphite substrates.

The results presented here have significant implications for future research on the chemical and physical properties of gold-based nanoparticles and should be of interest to many researchers working on experimental and theoretical aspects of nanoscience.

Edgbaston, April 2013 Prof. Roy L. Johnston

Acknowledgments

The work presented in this thesis would have been neither as interesting nor as satisfying had it not been for the company and support of those around me over the past three years. To all of you I am truly grateful.

I am wholly indebted to my supervisor Prof. Roy Johnston for his knowledge, dedication, enthusiasm, time and support; without whom the guidance of my work would have been a very much different, and perhaps more onerous, task. I am also beholden to Dr. Jaakko Akola, Dr. Ziyou Li, Prof. Jun Yuan and Dr. Sarah Horswell whose professional manner and academic insight has progressed my work over the recent years through our collaborative links.

I am most thankful to Dr. Benjamin Curley, Dr. Lauro Oliver Paz-Borbon and Dr. Andrew Bennett for many supporting conversations and educated contributions on the topic of computer programming, computational modelling and cluster science, and to Dr. Zhiwei Wang and Ruth Chantry for conversations on electron microscopy.

Financial support has been gratefully received from the EPSRC, UK (DTA Award Reference: EP/P504678/1) and the University of Birmingham, UK. Trans-European academic visits have been supported by the HPC-EUROPA2 project (project number: 228398), with the support of the European Community—Research Infrastructure Action of the FP7.

Contents

Abbreviations and Acronyms

2D	Two-dimensional
3D	Three-dimensional
AR	Aspect ratio
DDA	Discrete dipole approximation
DFT	Density functional theory
D-GA	Darwinian genetic algorithm
Dh	Decahedron
DOS	Density of states
ELF	Electron localisation function
FCC	Face-centred cubic
FE	Function evaluation
FEN	First encounter
FWHM	Full width, half maximum
GA	Genetic algorithm
GM	Global minimum
GPAW	Grid projector augmented wave
GR2	Graphite system, represented by two graphene layers
GR2-h	Graphite system, represented by two graphene layers, with a surface defect
HAADF	High-angle annular dark field
HE	Historical evaluation
HOPG	Highly ordered pyrolytic graphite
HOMO	Highest occupied molecular orbital
Ih	Icosahedron
I-Dh	Ino-decahedron
L-GA	Lamarckian genetic algorithm
LSF	Least squares fitting
LUMO	Lowest unoccupied molecular orbital
MP	Monolayer-protected
M-Dh	Marks-decahedron
MPI	Message passing interface
NC	Nanocluster
NP	Nanoparticle

OM	Offspring mutation
OMP	Open message passing
PAW	Projector augmented wave
PLDOS	Projected local density of states
PM	Parent mutation
SPR	Surface plasmon resonance
STEM	Scanning transmission electron microscopy
STM	Scanning tunnelling microscopy
XRD	X-ray diffraction

Chapter 1
Introduction

1.1 Nanoscience

Nanoscience, the scientific study of objects on the nanometre regime, has been a rapidly developing area of the physical sciences in the past 20 years due to both the commercial potential and academic curiosity with which scientists view this field. Nanoscience represents a crossover area between the bulk and atomic properties of a material, and the potential for discovering properties distinct from these two extremes is one of the major driving factors in continued investigation.

Historically, nanoscience is an under-studied field due to the experimental difficulty of controlled synthesis and study of particles on the nanoscale. The precision of analytical equipment has increased dramatically with the advent of modern electronics, thus creating new possibilities for the depth at which samples can be studied. Good examples include scanning transmission electron microscopy (STEM) and scanning tunnelling microscopy (STM), where sample manipulation [1] and imaging resolution [2, 3] is now possible on the atomic scale. The aforementioned improvement in electronic components has also led to the rapid expansion of computer modelling: it is now seen as a necessary accompaniment to most areas of scientific research [4]. Whereas experimental work has proceeded in a "top down" approach, working downwards from bulk configurations, computational simulations have developed in an opposing "bottom up" direction, progressing from atomic calculations upwards as the computing power available has increased rapidly over time. The eventual crossover between computational and experimental work has only come to fruition in the last 15–20 years, and nanoscientific research has been a major benefactor of these improvements.

The potential applications of nanoscience in technology makes it an exciting area for industrial and commercial expansion. For particles on the nanoscale, or "nanoparticles" (NPs), the multi-disciplinary application to physics, chemistry, biology, medicine and beyond are one of the driving forces behind continued capital investment. Most prominent in our everyday lives are the applications to electronics, where the public demand for higher-specification computational capabilities has pushed

A. J. Logsdail, *Computational Characterisation of Gold Nanocluster Structures*, Springer Theses, DOI: 10.1007/978-3-319-01493-7_1,

technology to the edge of classical physics, at which point new alternatives must be investigated beyond the nanoscale in the quantum regime e.g. quantum computation [5]. Another factor with a high-impact to the quality of our lives is the application of nanoscience to catalysis, or "nanocatalysis"[6]. The high surface to volume ratio of NPs not only makes them suitable for catalysis [7], it also offers chemical properties unlike the bulk phase [8], allowing for application to previously unavailable reaction pathways. Thirdly, the modern identification of novel optical properties in metallic nanoparticles has led to an up-surge in applications in the field of medicine, or "nanomedicine", for the site-specific delivery of drugs and identification of differing organic compounds [9].

1.2 Clusters

Clusters on the nanoscale, or "nanoclusters" (NCs), are groups of between a few and many millions of atoms or molecules. The distinction between clusters and bulk materials is normally distinguished by the ability to count the atoms in a cluster; NCs provide a link between bulk materials and isolated atoms, and they can often display distinct physical and chemical properties due to their varying size and composition. Clusters can consist of either identical atoms or molecules, or two or more different species. Clusters are formed by most of the elements of the periodic table, and can be studied in a variety of media [7, 10]. They have wide-ranging applications in the fields of catalysis, optics and magnetism [6, 11–13]. Knowledge of the geometrical, electronic and compositional influences is key in understanding the peculiar characteristics NCs display [10, 14, 15].

Physical theory provides a useful accompaniment for experimental observations, and, as mentioned above, the advent of computers has allowed calculation performance to improve dramatically over recent years. The prediction of a variety of aforementioned properties is possible; one particularly important tool to the computational chemist being density functional theory (DFT) [16], which is a first-principle, or *ab initio*, method for the calculation of the electronic structure of atoms, NCs and bulk materials, from which further properties can be derived. Due to computational expense, the forecasting of NC properties with DFT have been mostly limited to tens of atoms, with calculations being performed in the gas-phase. Experimentally, gas-phase measurements can be achieved using methods such as mass spectroscopy (MS) and photoelectron spectroscopy (PES), with notable scientific breakthroughs being the identification of the extremely stable C_{60} fullerene [17], and also the Au_{20} tetrahedron [18]. The properties of small NCs, such as ionisation potential, the gap between the highest occupied molecular orbital (HOMO) and lowest unoccupied molecular orbital (LUMO), and reactivity, often show a pronounced size dependence [7].

The extended experimental study of large metallic NCs in the gas-phase proves difficult due to the difficulty in entrapping NCs and further preventing their agglomeration under reactive conditions. These limitations, and the time-consuming nature of gas-phase study for large NCs, mean that the stabilisation of clusters by a supportive medium is increasingly favourable in experimental work [15]; be it deposited on a substrate surface [19], suspended in a solution [20] or ligand stabilised in a crystallised form [21]. Clusters can then be studied (repeatedly) using STEM, X-ray diffraction (XRD) or UV-Vis spectroscopy, to name but a few methods. Of course, the supports can further effect the structural and electronic properties of the NCs [19], which will effect the chemical functionality of the NCs [22, 23]. Computationally these extra considerations are a requirement of modern analysis, though the representation of the supportive medium can vastly increase the computing requirements for simulation.

1.2.1 Geometric Size Effects

Recent experimental and theoretical work has shown transitions-metal NCs to present both crystalline structures (e.g. FCC), akin to bulk materials, and also non-trivial non-crystalline structures [3, 15, 24–26]. Overall, clusters of size N favour structures that minimise total potential energy [7] at low temperatures; the ideal binding energy (E_b) of a cluster of size N can be written in the form [27, 28]:

$$E_b = aN + bN^{\frac{2}{3}} + cN^{\frac{1}{3}} + d \quad (1.1)$$

where the first term represents volume contribution (aN), followed by surface contributions from facets ($bN^{\frac{2}{3}}$), edges ($cN^{\frac{1}{3}}$) and vertices (d) respectively. Adopted structures are very much dependent on N: the stability of different geometric structure-types depends on the relationship between the exposed surface area, and surface energy of the various exposed faces, against the internal strain caused by atomic rearrangement with respect to the bulk crystal lattice [24].

It is easy to mathematically approximate the relationship between the number of atoms, N, in a NC and the percentage of those atoms which occupy a surface site ($N^{\frac{1}{3}}$) using the liquid drop model (LDM) [7], however the geometric shape can also play a major part in the structure of a cluster. "Magic number" is a term used to refer to values of N where a complete shell is formed of a geometric polyhedra by the atoms [15, 29]. The complete shells result in an unusually high-level of stability with respect to the structures found for surrounding nuclearities, and have been documented for rare-gas, organic, alkali-metal and other metallic clusters [7].

At smaller NC sizes non-crystalline structures are more common, and as N increases the structures tend towards fragments of the crystalline bulk arrangement [7, 24]—with some "magic numbers" along the way [24, 30, 31]. Common small cluster structures of non-crystalline nature include the Mackay icosahedra (Ih) [32], the Ino-decahedra (I-Dh) [33] and the Marks truncated decahedra (M-Dh) [34].

1.2.2 Electronic Size Effects

The electronic closure of shells within a cluster was first noted Knight et al. for sodium clusters up to 100 atoms [35], where it was noted that closed atomic shells of delocalised valence electrons led to increased stability in specific cases: $N = 8$, 20, 40, 58 and 92 (and thus an equivalent number of valence electrons). The electronic structure was modelled by applying the Jellium, or uniform electron gas (UEG), model [7] to the scenario, whereby an electron gas is modelled moving in an ionic pseudo-potential from the bulk, and averaged to a spherically symmetric potential well. Solving the Schrödinger equation for each value of N revealed discrete states with decreasing energies and degeneracy. Since this discovery "superatoms" have also been identified for other alkali metals [29], and metal clusters such as aluminium [36] and gold, both homogeneously [18] and in more complex structures [37]. The specific number of valence electrons which cause shell-closure—8, 20, 40, etc.—are also termed "magic number" sizes within this model [7, 15].

Another characteristic electronic feature of transition metals is their ability to conduct in the bulk phase. This conductivity is due to continuous electron bands, which allow the flow of electrons, and more specifically d-orbital electrons, across a material. As we reduce the number of atoms from the bulk phase into the regime of small clusters, these energy levels become discrete, turning the metallic clusters into insulators; this is known as the size-induced metal-insulator transition [7], and can be quantified by the size of the Kubo gap (δ) at the Fermi Energy (E_f) [38]:

$$\delta = \frac{4E_f}{3N} \tag{1.2}$$

where N is the number of valence electrons. If the gap is smaller than the thermal energy, k_BT, the particle will be metallic, and if the gap is larger than k_BT then the particle will be insulating.

In medium to large metal NPs, and in bulk metals, the collective oscillation of delocalised electrons at the interface between the metal and its surrounding medium (i.e. surfaces and edges) readily occurs, and such resonances are known as surface plasmons. For NPs where the wavelength of the light is significantly larger than the longest NP dimension, these oscillations become coupled with photons in what is called a localised surface plasmon resonance (LSPR) [39, 40]. For nanoparticles adsorption dominates the spectral field, whilst scattering is more prominent in larger particles. LSPRs exhibit enhanced near-field amplitude, which decays rapidly away from the NP, at the resonance wavelength [41]. LSPRs have been used to greatly enhance several spectroscopic techniques, including fluorescence and Raman scattering [38, 42]. Commercially, potential use has been shown for improving photovoltaic cells by depositing metal nanoparticles on their surface [43].

1.2.3 Compositional Effects

In materials chemistry, the range of properties displayed by a metallic system can be greatly improved by creating mixtures of elements to generate intermetallic compounds and alloys. The desire to fabricate materials with finely tuned, controllable properties and structures, on a nanometre scale, combined with the flexibility offered by alloy compounds, has created widespread interest in bimetallic alloy NCs, or "nanoalloys" [14, 44]. The study of their surface structures, compositions and segregation are of particular interest as they have a large impact on the chemical properties, and therefore the catalytic applications [14]. Nanoalloy segregation predominantly take one of four forms: core-shell segregated , sub-cluster segregated, multi-shelled and completely mixed [14].

From a theoretical perspective, the presence of more than one element also gives rise to a much larger variety of possible atomic configurations for each given geometry, as now each position can be occupied by one of two (or more) species. This leads to both geometric isomers and isomers based on permutations of unlike atoms. Jellinek et al. introduced the term "homotop" to describe A_aB_b nanoalloy isomers with a fixed number of atoms (N = a + b) and composition (a/b ratio), and the same geometrical arrangement of atoms, but which differ in the way the atom types are arranged [45]. A single geometrical isomer of an N-atom A_aB_b cluster will give ${}^NP_{a,b}$ homotops, where [14]:

$$
{}^NP_{a,b} = \frac{N!}{N_a!N_b!} = \frac{N!}{N_a!(N-N_a)!} \tag{1.3}
$$

As N increases so ${}^NP_{a,b}$ rapidly rises, complicating the identification of compositional minima. Approximations to the segregation can be used for larger NCs, as experimentally one can identify segregated compositional isomers using techniques such as UV-Vis spectra and STEM [46], and we will be discussing this further.

1.3 This Work

In this research we look to investigate and improve the structural characterisation of metallic nanoclusters, coupling the experimental results of others with our theoretical calculations. Both homogeneous and heterogeneous clusters are studied, with the common theme throughout being the presence of gold. Gold is of particular interest due to the catalytic properties it displays in nanocluster form [8], a feature uncommon to the bulk form, as well as other potential uses in optical and electronic technology. However, gold is expensive, and so bimetallic derivatives are also investigated in the search for equivalent properties and/or novel new features.

Starting with Chap. 2, structural motif preference is investigated as a function of composition and size for Au and Pd, concentrating on high-symmetry structures. 12-vertex bimetallic $(AuPd)_N$ core/shell particles are studied further, both $Au_{core}Pd_{shell}$ and the inverse $Pd_{core}Au_{shell}$ arrangements. We aim to identify the relative structural preferences of both the monometallic and bimetallic clusters.

Chapter 3 discusses the development and implementation of a STEM model and coupling this with a genetic algorithm, designed to aid the determination of the structures of small nanoclusters from experimental STEM images. We would like to implement a search method which is efficient and accurate at identifying experimental nanocluster images so as to help corroborate our computational results with experimental observations. We discuss development of both the STEM model and search methods, with the objective of identifying both promising characterisations in our own work and potential future improvements.

Chapter 4 contains a first-principles study of Au_{16} deposited on a graphite substrate. Surface supported clusters are of catalytic interest, and thus we investigate the stability of the Au_{16} in the presence of a variety of defect sizes. We hope to elucidate the electronic properties of the adsorbed clusters, and we culminate by looking at the potential use of the adsorbed Au_{16} clusters as a catalyst in the oxidation of carbon monoxide.

Finally, Chap. 5 investigates the optical response of monometallic and bimetallic nanoparticles. The aim of our work in this section is to look at what effects the optical spectra: nanoparticle geometry and structural composition are both used as "tunable" parameters. The optical response of Au is very characterisable in the UV-Visible region, and we test various elongated bimetallic particles with an aim to finding out more information on the effect bimetallic compositions have on the absorption spectra.

Each chapter in this manuscript is self-containing, with a short introduction, methodology, conclusions and bibliography given along with detailed accounts of the observations witnessed for each section. Overall concluding remarks are given at the end.

References

1. D.M. Eigler, E.K. Schweizer, Positioning single atoms with a scanning tunnelling microscope. Nature **344**(6266), (1990)
2. N.D. Browning, M.F. Chisholm, S.J. Pennycook, Atomic-resolution chemical analysis using a scanning transmission electron microscope. Nature **366**(6451), 143–146 (1993)
3. Z.Y. Li, N.P. Young, M. Di Vece, S. Palomba, R.E. Palmer, A.L. Bleloch, B.C. Curley, R.L. Johnston, J. Jiang, J. Yuan, Three-dimensional atomic-scale structure of size-selected gold nanoclusters. Nature **451**, 46–48 (2008)
4. A. Hinchliffe, *Chemical Modelling: From Atoms to Liquids* (Wiley, Chichester, 1999)
5. A. Steane, Quantum computing. Rep. Prog. Phys. **61**(2), 117 (1998)
6. U. Heiz, U. Landman, *Nanocatalysis* (Springer-Verlag, Berlin, 2007)
7. R.L. Johnston, *Atomic and Molecular Clusters* (Taylor and Francis, London, 2002)
8. M. Haruta, S. Tsubota, T. Kobayashi, H. Kageyama, M.J. Genet, B. Delmon, Low-temperature oxidation of CO over gold supported on $TiO_2, \alpha-Fe_2O_3$, and Co_3O_4. J. Catal. **144**(1), 175–192 (1993)

9. T.M. Allen, P.R. Cullis, Drug delivery systems: entering the mainstream. Science **303**(5665), 1818–1822 (2004)
10. H. Haberland, *Clusters of Atoms and Molecules* (Springer-Verlag, Berlin, 1994)
11. P.C. Ray, Size and shape dependent second order nonlinear optical properties of nanomaterials and their application in biological and chemical sensing. Chem. Rev. **110**(9), 5332–5365 (2010)
12. J.P. Liu, E. Fullerton, O. Gutfleisch, D.J. Sellmyer, *Nanoscale Magnetic Materials and Applications* (Springer, New York, 2009)
13. B. Corain, G. Shemid, N. Toshima, *Metal Nanoclusters in Catalysis and Materials Science: The Issue of Size Control* (Elsevier, Amsterdam, 2008)
14. R. Ferrando, J. Jellinek, R.L. Johnston, Nanoalloys: from theory to applications of alloy clusters and nanoparticles. Chem. Rev. **108**(3), 845–910 (2008)
15. F. Baletto, R. Ferrando, Structural properties of nanoclusters: energetic, thermodynamic, and kinetic effects. Rev. Mod. Phys. **77**, 371–423 (2005)
16. P. Hohenberg, W. Kohn, Inhomogeneous electron gas. Phys. Rev. **136**, B864–B871 (1964)
17. H.W. Kroto, J.R. Heath, S.C. Obrien, R.F. Curl, R.E. Smalley, C(60): buckminsterfullerene. Nature **318**, 162–163 (1985)
18. J. Li, X. Li, H.J. Zhai, L.S. Wang, Au_{20}: a tetrahedral cluster. Science **299**, 864–867 (2003)
19. S.J. Carroll, in *Deposition of Size-selected Atomic Clusters on Surfaces*, Ph.D. thesis, School of Physics and Astronomy (University of Birmingham, Birmingham, 1999)
20. J. Pérez-Juste, I. Pastoriza-Santos, L.M. Liz-Marzán, P. Mulvaney, Gold nanorods: synthesis, characterization and applications. Coord. Chem. Rev. **249**(17–18), 1870–1901 (2005)
21. M.M. Alvarez, J.T. Khoury, T.G. Schaaff, M.N. Shafigullin, I. Vezmar, R.L. Whetten, Optical absorption spectra of nanocrystal gold molecules. J. Phys. Chem. B **101**(19), 3706–3712 (1997)
22. M. Valden, X. Lai, D.W. Goodman, Onset of catalytic activity of gold clusters on titania with the appearance of nonmetallic properties. Science **281**(5383), 1647–1650 (1998)
23. S. Lee, B. Lee, F. Mehmood, S. Seifert, J.A. Libera, J.W. Elam, J. Greeley, P. Zapol, L.A. Curtiss, M.J. Pellin, P.C. Stair, R.E. Winans, S. Vajda, Oxidative decomposition of methanol on subnanometer palladium clusters: the effect of catalyst size and support composition. J. Phys. Chem. C **114**, 10342–10348 (2010)
24. F. Baletto, R. Ferrando, A. Fortunelli, F. Montalenti, C. Mottet, Crossover among structural motifs in transition and noble-metal clusters. J. Chem. Phys. **116**(9), 3856–3863 (2002)
25. R. Ferrando, A. Fortunelli, R.L. Johnston, Searching for the optimum structures of alloy nanoclusters. Chem. Phys. Phys. Chem. **10**, 640–649 (2008)
26. L.O. Paz-Borbón, R.L. Johnston, G. Barcaro, A. Fortunelli, Structural motifs, mixing, and segregation effects in 38-atom binary clusters. J. Chem. Phys. **128**(13), 134517 (2008)
27. C.L. Cleveland, U. Landman, The energetics and structure of nickel clusters: size dependence. J. Chem. Phys. **94**(11), 7376–7396 (1991)
28. J. Uppenbrink, D.J. Wales, Structure and energetics of model metal clusters. J. Chem. Phys. **96**(11), 8520–8534 (1992)
29. T.P. Martin, Shells of atoms. Phys. Rep. **273**, 199–241 (1996)
30. J.P.K. Doye, D.J. Wales, Chem. Phys. Lett. **247**, 339–347 (1995)
31. L.O. Paz-Borbón, T.V. Mortimer-Jones, R.L. Johnston, A. Posada-Amarillas, G. Barcaro, A. Fortunelli, Structures and energetics of 98 atom Pd Pt nanoalloys: potential stability of the Leary tetrahedron for bimetallic nanoparticles. Phys. Chem. Chem. Phys. **9**, 5202–5208 (2007)
32. A.L. Mackay, A dense non-crystallographic packing of equal spheres. Acta. Cryst. **15**(9), 916–918 (1962)
33. S. Ino, Stability of multiply-twinned particles. J. Phys. Soc. Jpn. **27**(4), 941–953 (1969)
34. L.D. Marks, Surface structure and energetics of multiply twinned particles. Philos. Mag. A **49**, 81–93 (1984)
35. W.D. Knight, K. Clemenger, W.A. de Heer, W.A. Saunders, M.Y. Chou, M.L. Cohen, Electronic shell structure and abundances of sodium clusters. Phys. Rev. Lett. **52**(24), 2141–2143 (1984)
36. D.E. Bergeron, P.J. Roach, A.W. Castleman, N.O. Jones, S.N. Khanna, Al cluster superatoms as halogens in polyhalides and as alkaline earths in iodide salts. Science **307**(5707), 231–235 (2005)

37. M. Walter, J. Akola, O. Lopez-Acevedo, P.D. Jadzinsky, G. Calero, C.J. Ackerson, R.L. Whetten, H. Grönbeck, H. Häkkinen, A unified view of ligand-protected gold clusters as superatom complexes. Proc. Natl. Acad. Sci. **105**, 9157–9162 (2008)
38. S.M. Morton, D.W. Silverstein, L. Jensen, Theoretical studies of plasmonics using electronic structure methods. Chem. Rev. **111**, 3962–3994 (2011)
39. R.H. Ritchie, Plasma losses by fast electrons in thin films. Phys. Rev. **106**, 874–881 (1957)
40. E.A. Stern, R.A. Ferrell, Surface plasma oscillations of a degenerate electron gas. Phys. Rev. **120**(1), 130–136 (1960)
41. E. Hutter, J.H. Fendler, Exploitation of localized surface plasmon resonance. Advanced Materials **16**(19), 1685–1706 (2004)
42. M.E. Stewart, C.R. Anderton, L.B. Thompson, J. Maria, S.K. Gray, J.A. Rogers, R.G. Nuzzo, Nanostructured plasmonic sensors. Chem. Rev. **108**, 494–521 (2008)
43. U. Kreibig, M. Vollmer, *Optical Properties of Metal Clusters* (Springer, Berlin, 1995)
44. Nanoalloys: From Theory to Applications. In *Faraday Discussions No.138*, vol. 138, (University of Birmingham, Birmingham, 2008)
45. J. Jellinek, E.B. Krissinel, Ni_nAl_m alloy clusters: analysis of structural forms and their energy ordering. Chem. Phys. Lett. **258**(1–2), 283–292 (1996)
46. Z.Y. Li, J.P. Wilcoxon, F. Yin, Y. Chen, R.E. Palmer, R.L. Johnston, Structures and Optical Properties of 4–5 nm Bimetallic AgAu Nanoparticles, in *Faraday Discussions*, vol. 138, (University of Birmingham, Birmingham, 2008), p. 363

Chapter 2
Calculating the Structural Preference of High Symmetry Clusters for Pd_N, Au_N, and $(PdAu)_N$

2.1 Introduction

Computational modelling plays an important role in understanding the properties of nanoclusters, as it allows the prediction of structures for the lowest energy isomer (i.e. the global minimum, GM) [1], as well as providing information on preferential cluster geometries [2], or local minima, and further details such as metal segregation in bimetallic systems [3]. In this chapter the relative energetics will be discussed for different high-symmetry structures composed of Pd, Au, or a combination of the two. Clusters have been created using mathematical constructs, and then energetically minimised. Stability trends are identified for different compositions and geometries, in order to compare our results with experimental observations.

A large amount of computational research work has been done previously on both homogeneous and heterogeneous clusters; recent comprehensive reviews have been carried out by Baletto et al. [4] and Ferrando et al. [3] respectively, and we highlight some of the key publications below.

2.1.1 Pd_N Clusters

The energetics of small Pd nanoparticles have been thoroughly investigated by several computational groups using first-principle techniques [2, 5–10]. Besides their catalytic properties, work on these clusters is interesting as their lowest-lying isomer may have nonzero spin [11], thus being magnetic. At large sizes *ab initio* calculations are rare due to their computational expense: DFT calculations by Nava et al. found an early preference for bulk-like structures when the number of atoms $N > 100$, with a larger binding energy (E_b) for FCC geometries than icosahedra (Ih) or Ino-decahedra (I-Dh) [5]—however at "magic numbers" Ih and cuboctahedra (CO) remain very energetically competitive; a result confirmed by Guirado-López et al. [9]. Work by Baletto and Ferrando using the RGL potential [12, 13] on large

A. J. Logsdail, *Computational Characterisation of Gold Nanocluster Structures*,
Springer Theses, DOI: 10.1007/978-3-319-01493-7_2,

Pd_N clusters ($N < 5{,}000$) reviewed relative geometric stabilities over a wide range of sizes, showing a preferential trend of Ih for low nuclearities, FCC for large nuclearities and M-Dh in between, with interval windows greater than observed in a similar study of Au_N clusters [2].

In experiments, José-Yacamán et al. observed thiol-passivated Pd nanoparticles by TEM, witnessing structures ranging from crystalline (FCC) to high-symmetry (Ih); very large Ih were also observed, their presence associated with kinetic trapping effects, and these results were backed up with theoretical images [14]. A review of other recent work on Pd nanoclusters by José-Yacamán and co-workers reaffirms the presence of a variety of structures, and also mentions the influence the substrate may have upon structural growth [15]. Size-selected Pd_N clusters were studied by Pearmain [16] using scanning transmission electron microscopy (STEM). A mixture of Ih, I-Dh and FCC structures were identified for $N = 887$, as well as structural changes for larger nuclearities used in catalytic processes.

2.1.2 Au_N Clusters

Small Au_N clusters ($N \leq 10$) have been thoroughly investigated using *ab initio* techniques, showing a preference for planar geometries [17–19]. DFT calculations and IR photodepletion experiments have shown that Au_{20} is a tetrahedron [20, 21], with FCC packing; for other sizes low-symmetry structures are known to be energetically competitive [22]. For larger clusters *ab initio* calculations are less feasible: as previously mentioned Baletto and Ferrando carried out calculations on high-symmetry structures using the RGL potential, where a preferential structural interval for Ih was identified for $0 < N \lesssim 50$, followed by a M-Dh interval of similar size ($50 \lesssim N \lesssim 100$) [2]. After this crystalline structures were prominent, and similar results were found by various other groups using other semi-empirical potentials [23, 24].

Experimental studies of Au_N clusters are difficult as they are often passivated after formation, and it is difficult to know what effect the support agents have [25]. Recently Häkkinen, Akola and co-workers performed geometry calculations for thiolate-supported of Au_{25}, Au_{38}, Au_{102}, and Au_{144} systems, as well as other phosphine-halide stabilised Au nanoparticles in the same size-range using DFT and matching their results to experimental X-ray diffraction patterns [26–31]. The calculations indicate that small ($\leq$2 nm) passivated Au nanoparticles can be viewed as electronically stable superatoms with interesting atomic rearrangements at the interface between the metallic core and ligand-shell.

Experimentally, Wang et al. looked at characterising the thiolate-supported Au_{38} clusters using STEM, however the emphasis of this work was on atom counting as the low-symmetry of small clusters makes identifying motifs difficult [32]. Li, Palmer, Johnston and co-workers [33, 34] identified Au_{309} cluster geometries by comparing experimental HAADF-STEM images with image simulations (based on low energy structures found using the semi-empirical Gupta potential), and found a number of competing structures; Ih, I-Dh and crystalline (FCC-like) geometries were all observed. STEM characterisation of Au_{887} is of current interest to this group of collaborators, for comparison with the previously mentioned Pd_{887} [35]. Stellated Au_N structures have been reported experimentally by the group of José-Yacamán [36], and Au_{887} is suspected to favour similar increased faceting.

2.1.3 $(PdAu)_N$ Clusters

Empirical calculations for small $(PdAu)_N$ clusters have been conducted by the Johnston group using the Gupta Potential coupled with a genetic algorithm (GA) [37] for $N < 50$ [38], and also specifically $N = 34$ and 38 [39]. In both these works various potential parameter sets were tested, with parameters first derived by Cleri and Rosato [12] proving to favour $Pd_{core}Au_{shell}$ segregation, whilst more recent parameters derived from *ab initio* calculations indicated a more mixed configuration; higher level calculations were unable to offer any more conclusive opinions on the matter. These findings highlight the sensitivity of structural configurations on the parameter sets used, especially at small sizes [39].

Calculations for larger clusters using the embedded-atom model (EAM) showed preference for $Pd_{core}Au_{shell}$ segregation [40], which has been backed up by classical molecular dynamics (MD) calculations indicating core/shell inversion from $Au_{core}Pd_{shell}$ to $Pd_{core}Au_{shell}$ above 500 K [41]; the lower surface energy (E_{surf}) of gold (Au: 96.8 meV Å^{-2}, Pd: 131 meV Å^{-2} [42]) is thought to be the driving force behind this configuration. Experimental work conducted in the same article reports the successful synthesis of $Au_{core}Pd_{shell}$, $Pd_{core}Au_{shell}$ and mixed PdAu nanoparticles [41], showing thermodynamic driving forces may be restricted by kinetic trapping depending on the synthetic method. In larger clusters, Ferrer et al. have reported the experimental formation of tri-layered Au/Pd nanoparticles [43].

2.2 Methodology

2.2.1 High Symmetry Structures

It has been shown that three common high-symmetry 12-vertex geometries, namely the Ih, I-Dh and CO, can be arranged from the same "magic number" nuclearities given by the third-order polynomial [44]:

$$N(k) = \frac{1}{3}[10k^3 + 15k^2 + 11k + 3] \quad (2.1)$$

Fig. 2.1 High symmetry structures with matching magic number nuclearities; illustrated for $k = 2$ ($N = 55$). From left to right: Ih, I-Dh and CO. Predominate faces are outlined in *solid red lines*, whilst background faces are marked with *dashed red lines*. Reproduced from [46] by permission of The Royal Society of Chemistry

where k is the number of shells in the structure—the first few values of N (i.e. "magic numbers") in this sequence are 13, 55, 147 and 309. Increases in k lead to a larger value of N but a lower proportion of surface atoms, which can be approximated as $N^{\frac{2}{3}}$.

Each of these identified structures (Ih, I-Dh and CO) have distinct atomic arrangements with respect to each other, leading to differing attributes (Fig. 2.1, left). The Ih consists of 20 (111) faces, making it the most spherical of the three structures. This arrangement minimises the surface area to volume ratio, whilst maximising nearest neighbours contact, at the expense of high strain energy. The Ih geometry is most common for small clusters, where the energetic cost of having large surfaces is too great unless exceptional electronic configurations exist [45]. For larger clusters the minimisation of internal strain energy is more important. The CO is a fragment of an FCC crystal, with 8 (111) faces and 6 (100 faces); strain energy is at its lowest due to the bulk crystal nature, and maintains a reasonable number of nearest-neighbour contacts (Fig. 2.1, right). The CO is most common for large clusters as we approach the bulk limit ($N \rightarrow \infty$). The I-Dh occupies the transitional space between these two geometries, with its structure not containing the tight packing of the Ih structure nor the high surface area of the CO: it has 10 (111) faces and 5 (100) faces, and is most commonly seen for medium sized clusters (Fig. 2.1, centre).

A structure closely related to the I-Dh is the Marks-Decahedron (M-Dh). The M-Dh has an I-Dh structure at its centre, and on to the (100) faces (Fig. 2.2, centre) protruding facets grow, relieving some of the internal strain at the cost of greater surface area. One can calculate the magic numbers for M-Dh clusters as [2]:

$$\begin{aligned} N(m,n,r) = \frac{1}{6}\{&30r^3 - 135r^2 + 207r - 102 \\ &+ [5m^3 + (30r - 45)m^2 + (60(r^2 - 3r) + 136)m] \\ &+ n[15m^2 + (60r - 75)m + 3(10r^2 - 30r) + 66]\} - 1 \end{aligned} \tag{2.2}$$

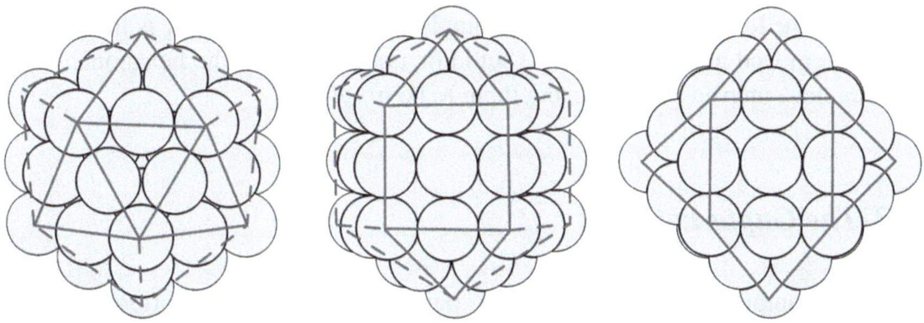

Fig. 2.2 As for Fig. 2.1, with structures rotated by 90° around an (x-)axis positioned horizontally across the page. Reproduced from [46] by permission of The Royal Society of Chemistry

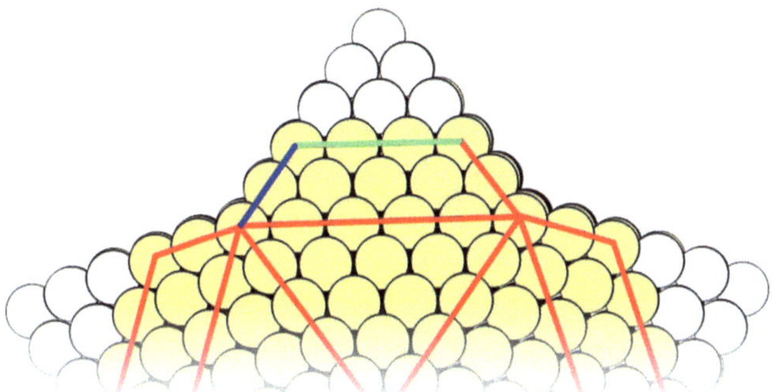

Fig. 2.3 Diagram of the re-entrant faceting on the M-Dh structure; illustrated for $(m, n, r) = (4, 4, 3)$. *Yellow spheres* represent atoms in our model, whilst *white spheres* show the limit of re-entrant facets ($r_{max} = 6$) on this size; this structure would be equivalent to setting $(m, n, r) = (1, 1, 6)$. *Solid red lines* define the faces on the $(4, 4, 3)$ structure, with m (*green*) and r (*blue*) highlighted

where m and n are the lengths of the sides of the (100) facets perpendicular and parallel to the fivefold axis, respectively, and r is the depth of the Marks-type re-entrant facets. For $r = 2$ and $m = n$, the magic numbers formed begin with the sequence $N = 18, 75, 192$ and 389. The maximum re-entrant depth (r_{max}) is limited by the length of the side on the I-Dh core, such that $r_{max} = k + 1$ (and when $m = n$, $r_{max} = m$) (Fig. 2.3).

Another high-symmetry structure often documented is the truncated octahedron (TO), a structure closely related to CO as both are fragments of the FCC crystal. The TO structure is created by the removal of the vertices from an octahedron to create a geometry with 6 (100) faces and 8 hexagonal (111) faces. We can define the number of atoms in the TO cluster [2]:

$$N(n_l, n_{cut}) = \frac{1}{3}(2n_l^3 + n_l) - 2n_{cut}^3 - 3n_{cut}^2 - n_{cut} \tag{2.3}$$

where n_l is the length of the edge of a complete octahedron and n_{cut} is the number of layers removed at each vertex. It should be noted here that the hexagonal (111) faces can have uneven sides, causing them to be irregular.

2.2.2 The Gupta Potential

It remains computationally exhaustive to perform *ab initio* calculations on large clusters ($N > 100$), and so empirical potentials remain widely used. Empirical potentials are derived from the fitting of experimental data to an assumed functional form, with common examples including the Murrell-Mottram [47], Gupta [12], Sutton-Chen [48] and EAM [49]. When coupled with efficient search techniques (e.g. GAs [37]) highly effective theoretical tools are formed.

The Gupta potential, derived from Gupta's expression for the cohesive energy (E_{coh}) of a bulk material [50], is based on the second moment approximation to tight-binding theory and is constructed from an attractive many-body (V^m) term and a repulsive pair (V^r) term, obtained by summing over all N atoms:

$$V_{clus} = \sum_{i=0}^{N}(V_i^r - V_i^m) \tag{2.4}$$

where

$$V_i^r(r_{ij}) = \sum_{j \neq i}^{N} A e^{-p(\frac{r_{ij}}{r_0}-1)} \tag{2.5}$$

and

$$V_i^m(r_{ij}) = \left[\sum_{j \neq i}^{N} \zeta^2 e^{-2q(\frac{r_{ij}}{r_0}-1)}\right]^{\frac{1}{2}} \tag{2.6}$$

In these equations, r_{ij} represents the distance between atoms i and j; and r_0 is the nearest neighbour distance in the bulk (in Å). The Gupta potential parameters (A, ζ, p and q) are fitted to the bulk properties of each metal (i.e. E_{coh}, bulk modulus, the annihilation of the energy gradient at r_0, and in some cases the surface energy, E_{surf}). After fitting to the bulk properties, one is left with only two independent parameters (p and q) that determine the range of the repulsive and attractive terms, respectively. The parameter sets I and II used in the following work are taken from the work of Cleri and Rosato [12] and Baletto et al. [2], respectively, and are shown in Tables 2.1 and 2.2, and plotted in Appendix A. Parameter sets are not mixed, however arithmetic means are used for bimetallic Au-Pd interactions.

Table 2.1 Gupta potential parameter sets I and II for Pd, taken from the work of Cleri and Rosato (*left*) and Baletto et al. (*right*), respectively [2, 12]

	Parameter Set I [12]	Parameter Set II [2]
A (eV)	0.1746	0.1715
ζ (eV)	1.718	1.702
p	10.867	11.00
q	3.742	3.79
r_0 (Å)	2.7485	2.7506

Table 2.2 Gupta potential parameter sets I and II for Au, taken from the work of Cleri and Rosato (*left*) and Baletto et al. (*right*), respectively [2, 12]

	Parameter Set I [12]	Parameter set II [2]
A (eV)	0.2061	0.2197
ζ (eV)	1.790	1.855
p	10.229	10.53
q	4.036	4.30
r_0 (Å)	2.884	2.878

2.2.3 *Long Range Cut-Off*

The parameters used in reference [2] incorporate a long distance cut-off into the potential for atoms further than 2 neighbours distance away ($>\sqrt{2}r$) [51]. The inter-atomic potential (Eq. 2.4) decays with increasing distance and introducing a cut-off speeds up the energy calculations for large clusters.

A 5th order polynomial was introduced, matching that of Baletto et al. [2]. For this, a polynomial replacement for the potential that matches the potential function at the cut-off start (C_s), and is zero at the cut-off end (C_e) is derived. The replacement is applied separately to each exponential component of the Gupta potential. Considering the term:

$$V_{ij}^r(r_{ij}) = Ae^{-p(\frac{r_{ij}}{r_0}-1)} \tag{2.7}$$

for $C_s \leq r_{ij} \leq C_e$ we replace the above expression by the following polynomial $p_1(r_{ij})$:

$$V_{ij}^r(r_{ij}) = p_1(r_{ij}) = a_5(r_{ij} - C_e)^5 + a_4(r_{ij} - C_e)^4 + a_3(r_{ij} - C_e)^3 \tag{2.8}$$

where the coefficients a_5, a_4, a_3 are chosen to match the function and its first and second derivatives for $r_{ij} = C_s$. For $r = C_e$, $p_1(r_{ij}) = 0$. The form of the polynomial ensures that the conditions on the function and its first derivative are automatically matched for $r_{ij} = C_e$. In the same way, for the term:

Table 2.3 Cut-offs and polynomial coefficients used for Au and Pd (to 4 significant figures) for parameter set *II*

	Parameter Set *II*	
	Pd	Au
C_s (Å)	3.890	4.070
C_e (Å)	4.764	4.984
a_3	-5.732×10^{-3}	-8.105×10^{-3}
a_4	-8.477×10^{-3}	-1.110×10^{-2}
a_5	-5.723×10^{-3}	-6.828×10^{-3}
x_3	−3.131	−2.232
x_4	−4.861	−3.258
x_5	−2.157	−1.383

$$\left[V_{ij}^{m}(r_{ij})\right]^{\frac{1}{2}} = \zeta e^{-q(\frac{r_{ij}}{r_0}-1)} \tag{2.9}$$

the following polynomial, $p_2(r_{ij})$, is used:

$$\left[V_{ij}^{m}(r_{ij})\right]^{\frac{1}{2}} = p_2(r_{ij}) = x_5(r_{ij} - C_e)^5 + x_4(r_{ij} - C_e)^4 + x_3(r_{ij} - C_e)^3 \tag{2.10}$$

where the coefficients x_5, x_4, x_3 are calculated to match the form of the Gupta function. The cut-off parameters used are given in Table 2.3, and included in the potential functions plotted in Appendix A.

2.2.4 Local Minimisation

The Limited-Memory Broyden-Fletcher-Goldfarb-Shanno algorithm (L-BFGS) algorithm [52], is a quasi-Newton method for finding the maxima and minima of a potential using a memory efficient gradient search. It computes the inverse Hessian ($\frac{d^2x}{dy^2}$) matrices of a function, but unlike in the traditional Newtonian method which has memory requirements of N^2 values, restricting the stored history to M steps results in a memory requirement of $N \times M$ values. This results in approximation of the Hessians, slowing down the convergence, but increases processing speed dramatically and thus speeds up the calculations, especially for large clusters ($N > 100$).

2.3 Energetic Analysis

In order to quantify our results, we must introduce several equations that allow us to compare cluster stability. Firstly we calculate the average binding energy per atom (E_b) using:

$$E_b = \frac{E_{tot}}{N} \tag{2.11}$$

where $E_{tot} = -V_{clus}$ is the total energy of the cluster. We can calculate relative stabilities (ΔE_b) of different geometries with identical nuclearities using:

$$\Delta E_b = E_{b_{Ih}} - E_{b_x} \tag{2.12}$$

where the binding energies of I-Dh or CO can be directly compared with Ih ($E_{b_{Ih}}$) when substituted in for E_{b_x}. Positive values indicate reduced stability, and a negative value indicates increased stability, relative to the Ih. The ratio of excess energy (E_{exc}) to the number of cluster surface atoms is given as [53, 54]:

$$\Delta = \frac{E_{exc}}{N^{\frac{2}{3}}} \tag{2.13}$$

and can be used as an indication of stability trends with size. E_{exc} is calculated as $E_{tot} - NE_{coh}$, where E_{coh} is the cohesive energy per atom in the bulk; thus $E_{exc} \propto E_b$. $N^{\frac{2}{3}}$ is used as an approximation to the number of surface atoms. In bimetallic systems E_{exc} becomes $E_{tot} - ME_{coh,a} - (N - M)E_{coh,b}$, where M is the number of atoms of type a, and $(N - M)$ is the number of atoms of type b. In general, we expect the following expression to hold for Δ [53, 54]:

$$\Delta = \frac{aN + bN^{\frac{1}{3}} + cN^{\frac{2}{3}} + d}{N^{\frac{2}{3}}} \tag{2.14}$$

The numerator is still equivalent to E_{exc}, and in this expression the constant d comes from the strain created by the vertices of the cluster, $cN^{\frac{1}{3}}$ represents strain from the edges, $bN^{\frac{2}{3}}$ from surface facets and the volume term aN is due to internal strain. aN vanishes for FCC geometries as $N \to \infty$, as it is the bulk crystal structure, whilst strain is present for all other geometries. Minimisation of Δ infers relative overall stability of a geometry.

2.4 Results and Discussion

High-symmetry structures were systematically created using the mathematical constructs outlined above, before being energetically minimised using the parameters of either Cleri and Rosato [12] or Balleto et al. [2], henceforth referred to exclusively as parameter sets I and II, respectively.

2.4.1 Pd_N Clusters

12-Vertex Structures: Icosahedra, Ino-Decahedra and Cuboctahedra

Figure 2.4 plots E_b for Ih, I-Dh and CO structures, with increasing shell sizes k, against N, using the Pd parameters given in Table 2.1. For $N < 100$, E_b rapidly increases, before levelling off as $N > 500$; this trend is consistent for both parameter sets. Identification between the different structural motifs is energetically difficult, implying that they are all closely competitive in energy at this size. A spacing between trend lines is visible for the two parameter sets: parameter set I rises to a higher level than parameter set II, before flattening out asymptotically relative to E_{coh}. The difference in gradients between parameter sets I and II can be identified in the right of Fig. 2.4 where an approximation to the E_{coh} is achieved for $N^{-\frac{1}{3}} \rightarrow 0$ when $N^{-\frac{1}{3}}$ is plotted against E_b. Linear extrapolation to 0 for parameter set I gives a slightly over-exaggerated E_{coh} (3.92 eV) compared to the experimentally measured value (3.89 eV [52]), whilst parameter set II offers better agreement to this measurement (3.88 eV).

Figure 2.5 plots ΔE_b in order to characterise the energetic preference of the Ih, I-Dh and CO structures in more detail. We can clearly see the close-packed Ih structure is the most stable for small sizes ($N < 1{,}000$). For the parameter set I, the linear intercept between $\Delta E_{b_{Ih}}$ and $\Delta E_{b_{I-Dh}}$ is at $N = 1{,}306$, whilst for parameter set II it is at $N = 1{,}068$. $\Delta E_{b_{CO}}$ remains positive throughout, and the intercept with $\Delta E_{b_{Ih}}$ would be for a value of $N > 1{,}500$. We would expect the intercept of $\Delta E_{b_{I-Dh}}$ and $\Delta E_{b_{CO}}$ to be at large values of N, as discussed in our preamble.

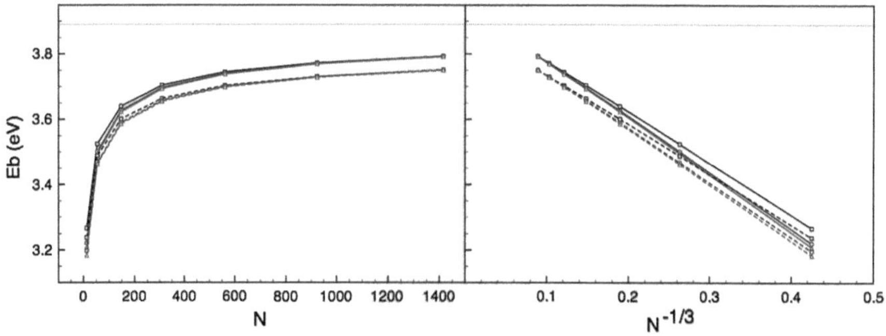

Fig. 2.4 *Left*: Plot of E_b for Pd_N structures against N. *Solid lines* represent the parameter set I of Cleri and Rosato [12], and *dashed lines* represent the parameter set II of Baletto et al. [2]. *Right*: Plot of E_b for Pd_N structures against $N^{-\frac{1}{3}}$. Ih (*black squares*), I-Dh (*red circles*) and CO (*blue triangles*) are shown in both plots; E_{coh} of 3.89 eV [52]) is displayed as a *grey solid line*

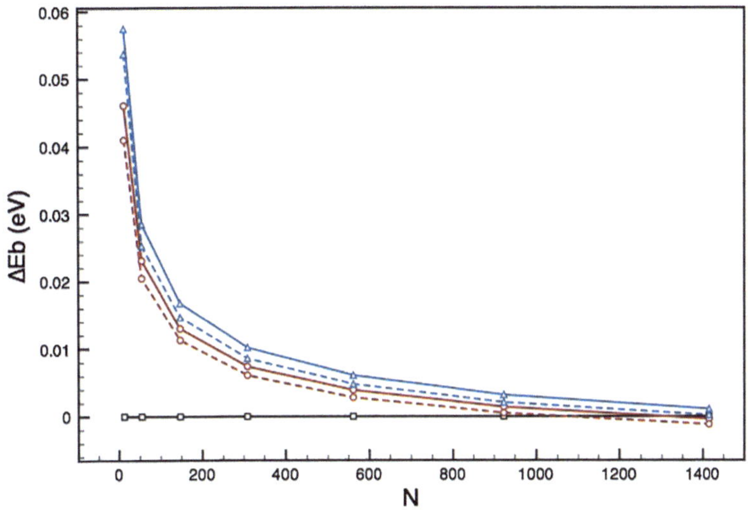

Fig. 2.5 Plot of ΔE_b for Ih (*black squares*), I-Dh (*red circles*) and CO (*blue triangles*) Pd_N structures against N. *Solid lines* represent the parameter set I of Cleri and Rosato [12], and *dashed lines* represent the parameter set II of Baletto et al. [2]

FCC Structures

FCC structures with a variety of shapes were compared to the CO structure (Fig. 2.6a). For the FCC structures we find a spread 1 eV $< \Delta <$ 2.2 eV. It is noticeable that the structures that minimise Δ for both the atom-centred and void-centred TO structures are closely competitive, with the adjoining lines difficult to distinguish. For both parameter sets I and II we find that the CO structure is uncompetitive with the lowest energy structures, which are TO, with Δ(TO) 0.1 eV less than Δ(CO) for similar N: detailed analysis shows the more stable TO structures as those with reduced (100) surfaces compared to the hexagonal (111)-packed faces. In Fig. 2.6b and c, TO structures for Pd_{314} and Pd_{861} are presented: in these cases, $n_l = 8$ and 11, respectively, and $n_{cut} = 2$.

Overall, Δ is similar for the parameters sets I and II when $N < 100$, however for $N > 500$ the minima of parameter set I are roughly 0.4 eV less than for parameter set II. Had we extended N further this gap would have increased, as the rate of change of the lowest values using parameter set I are considerably lower than that of parameter set II, though we would expect them to converge as $N \to \infty$.

Marks-Decahedra

Systematic variation of the M-Dh structure enables a detailed analysis of the effect of the signature re-entrant facets, for which Δ is plotted in Fig. 2.7. Trend lines are

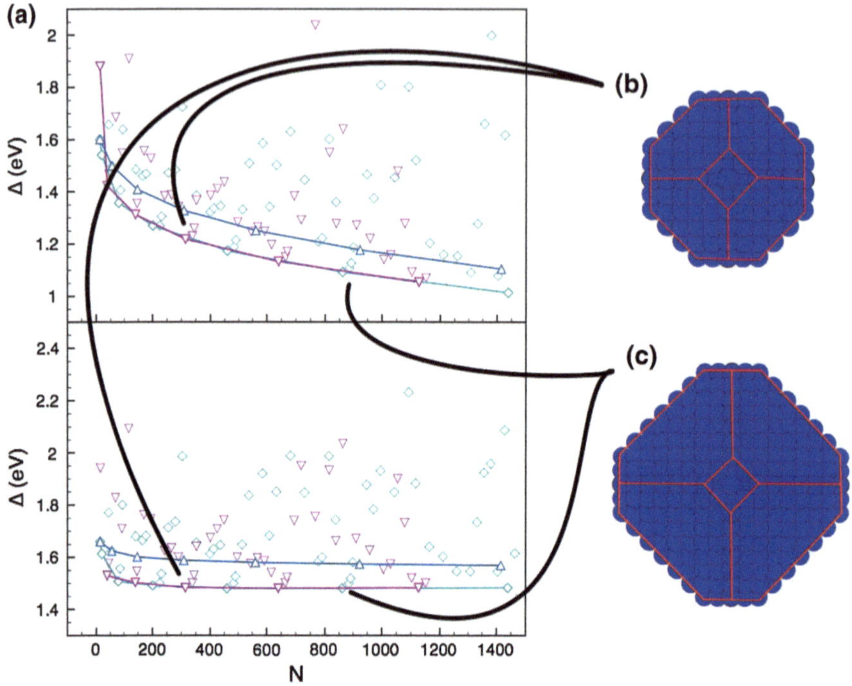

Fig. 2.6 (**a**) *Top*: Plot of Δ against N for various FCC Pd_N structures, where $N < 1{,}500$, for the parameter set I of Cleri and Rosato [12]. *Bottom*: As above, for the parameter set II of Baletto et al. [2]. TO structures with an atom at their centre (*green diamonds*) and with a void at their centre (*purple downward triangles*) are plotted against the CO structure (*blue upward triangles*) of increasing shell size k. *Lines* are plotted as a guide for the eye to the lowest values of Δ for each structure type. Highlighted structures are (**b**) Pd_{314}, a void-centred TO, with faces highlighted by *red lines* and (**c**) Pd_{861}, an atom-centred TO, with faces highlighted with *red lines*. Adapted from [55] by permission of the PCCP Owner Societies

identified for differing values of r_{max}, with their curved nature showing that the presence of truncated stellations is energetically favourable for both parameter sets I and II compared to the I-Dh structure, except at small sizes.

Parameter set I shows that the monolayer stellated structures ($r = 2$) (truncated or not) are most energetically favourable for $N < 600$, where stellations are truncated to a single layer of atoms on the I-Dh core. For $N > 600$ the ratio $\frac{r}{r_{max}}$ tends to a value of ~0.3, with multiple layers present on the truncated stellation. A progression in the trend lines occurs from sharp changes in Δ to shallower, more spaced out Δ values as N increases. The enlarged surface (100) sites on the I-Dh core (Fig. 2.2, centre) at large m and n mean even small re-entrant facets (e.g. $r = 2$) make large overall contributions to the value of N and thus the spacing of data points. The same observations hold true for the parameter set II, with $r = 2$ offering the lowest Δ values relative to structures with fuller stellations at $N < 300$, and as N increases so r does also, e.g. for $r_{max} = 9$ a truncation of the full stellated edges so $m = n = r = 5$

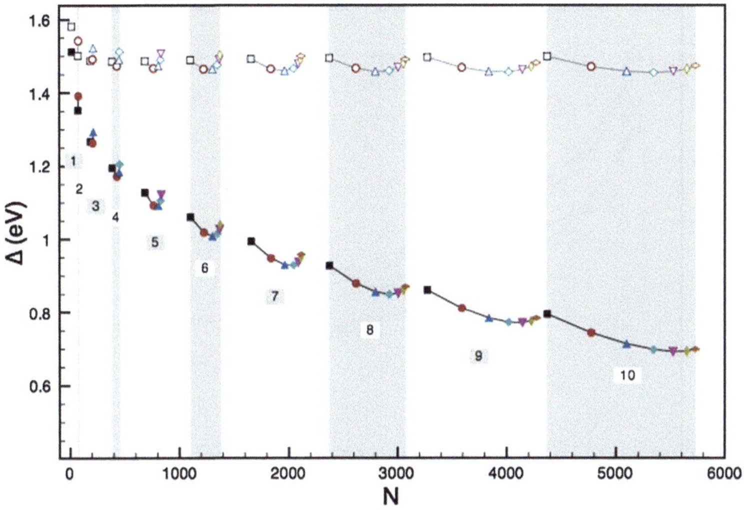

Fig. 2.7 Plot of Δ against N for Pd_N M-Dh of varying re-entrant facet sizes. Structures have been created such that $r < 9$, and $m = n < 11$ where $r = 2$ (*black squares*), 3 (*red circles*), 4 (*blue upward triangles*), 5 (*green diamonds*), 6 (*purple downward triangles*), 7 (*yellow laterally-extended diamonds*) and 8 (*brown vertically-extended diamonds*) are displayed. Parameter set I, of Cleri and Rosato [12], is plotted with closed symbols and *black trend lines*, and parameter set II, of Baletto et al. [2], is plotted with *open symbols* and *grey trend lines*. *Trend lines* illustrate the effect of reduced truncation of the stellated features from an I-Dh core ($r = 1$) to fully stellated structures ($m = 1$). Background colouring is added to highlight distinct trend lines. r_{max} is given below each line. Reproduced from [55] by permission of the PCCP Owner Societies

is preferred, a ratio value $\frac{r}{r_{max}} = 0.56$. The rate of overall change in Δ is decreased for parameter set II, similar to the observations made for TO fragments (Fig. 2.6).

Pd_{887}

A selection of the most energetically favourable Δ for Ih, I-Dh, CO, TO and M-Dh are plotted against N in Fig. 2.8, and compared to Pd_{887}. Pd_{887} is a cluster of recent experimental interest which is a magic number for M-Dh when $(m, n, r) = (4, 5, 3)$. As $m \neq n$ the (100) faces seen in the centre of Fig. 2.2 increase in size forming (110) faces, though still with (100) atomic packing (Fig. 2.9). For parameter set I the Pd_{887} cluster is 0.2 eV from the minimum, whilst for parameter set II this energy gap is much smaller (0.01 eV). In the experimental work of Pearmain [16], 86 % of 161 clusters were found to have irregular geometries, with small quantities of identifiable high-symmetry structures: 4 were identified as TO, whilst 5 more labelled as Ih, and only one as M-Dh. Whilst some of these limitations can be attributed to uncertainties in cluster orientation and experimental resolution limitations, the presence of several high-symmetry structures experimentally also indicates that these structures are close

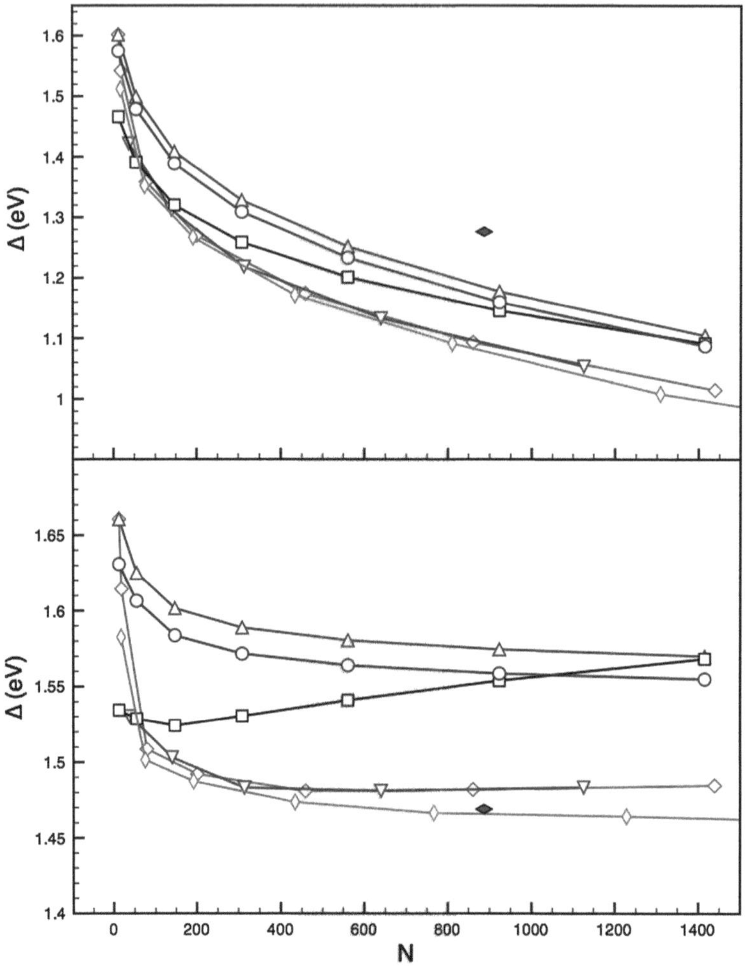

Fig. 2.8 *Top*: Plot of Δ for Pd_N structures against N, using parameter set I of Cleri and Rosato [12]. *Bottom*: As above, using parameter set II of Baletto et al. [2]. Ih (*black squares*), I-Dh (*red circles*), CO (*blue upward triangles*) are given, as well as the most energetically competitive atom-centred and void-centred TO (*purple downward triangles* and *green diamonds*, respectively), and M-Dh structures (*yellow vertical diamonds*). Pd_{887} is also shown (*brown filled lateral diamond*). Adapted from [55] by permission of the PCCP Owner Societies

in energy, and morphological changes under the electron beam are documented. Whilst the statistics provided from experiment is small, these experimental results fit to our calculations with parameter set I, where the energy gap to Pd_{887} is significantly large such that formation may only happen from kinetic trapping (not considering the impact the substrate may have on the structure).

Overall, Fig. 2.8 illustrates that parameter sets I and II have distinct energetic preference to the TO and M-Dh structures, compared to the Ih, I-Dh and CO. The Ih

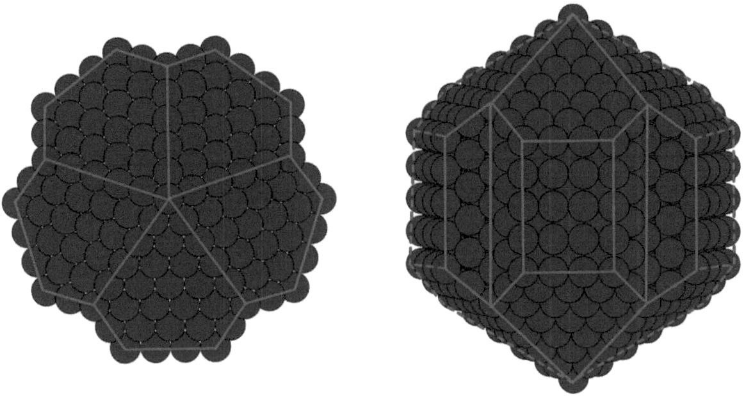

Fig. 2.9 Image of Pd_{887} M-Dh structure, where $(m, n, r) = (4, 5, 3)$. *Left*: View down five-fold axis. *Right*: View orthogonal to (*left*). Main faces are marked with *solid lines*

structure is competitive at low sizes ($N < 100$), before becoming less favourable as N increases. There is distinct energy differences between the trend lines for the high-symmetry 12-vertex Ih, I-Dh and CO structures, and the lower symmetry TO and M-Dh structures, with the latter proving more energetically favourable. However, for both parameter sets the energy difference for Δ is small between the TO/M-Dh trends and the 12-vertex Ih, I-Dh and CO structures (~0.1 eV), which equates to the thermal energy ($k_b T$) when $T \approx 1{,}200$ K, where T is temperature and k_b is the Boltzmann constant (8.617×10^{-5} eV K^{-1}). Though the energetic pathways between the different structures have not been calculated, we can be led to believe that transformation between these structures is not common under electron microscopy conditions, though subtler structural rearrangement have been observed [16].

2.4.2 Au_N Clusters

12-Vertex Structures: Icosahedra, Ino-Decahedra and Cuboctahedra

Figure 2.10 compares E_b for the high-symmetry Au_N structures of the Ih, I-Dh and CO geometries. Similar trends to the results of Pd_N are seen with sharply increasing E_b for $N < 100$, levelling out asymptotically towards the bulk E_{coh} value of 3.81 eV [52] as $N \rightarrow \infty$. Parameter set II tends to a higher value of E_b than parameter set I, an inverse of the results for Pd_N clusters (Fig. 2.4). Attention to the inset, which relates E_b to the bulk limit ($N^{-\frac{1}{3}} \rightarrow 0$), suggests that the parameter set I underestimates the extrapolated bulk E_{coh} value (3.77 eV), whilst parameter set II does not (3.81 eV).

Structural preferences are difficult to identify in Fig. 2.10; thus ΔE_b is plotted in Fig. 2.11. For parameter set I we see intersection of $E_{b_{Ih}}$ at $N = 682$ by $\Delta E_{b_{I-Dh}}$ and at $N = 923$ by $\Delta E_{b_{CO}}$, calculated using a linear fit. For parameter set II

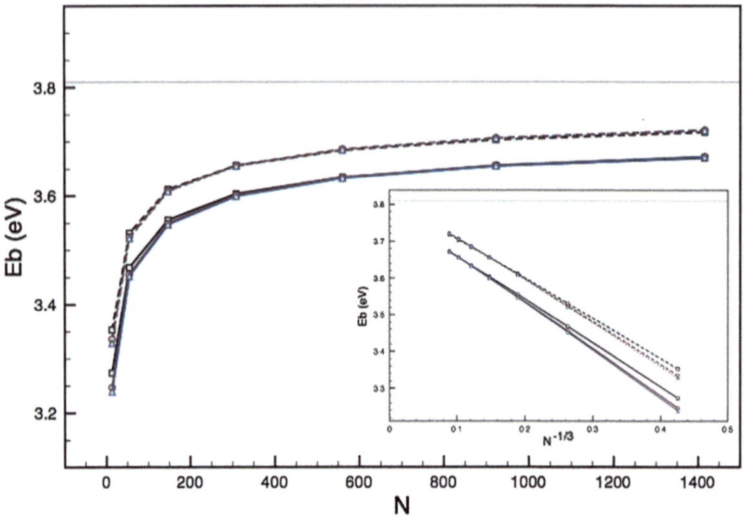

Fig. 2.10 *Main* Plot of E_b for Au_N structures against N. *Solid lines* represent the parameter set I of Cleri and Rosato [12], and *dashed lines* represent the parameter set II of Baletto et al. [2]. *Inset* Plot of E_b for Au_N structures against $N^{-\frac{1}{3}}$. Ih (*black squares*), I-Dh (*red circles*) and CO (*blue triangles*) are shown in both plots; the bulk E_{coh} (3.81 eV [52]) is displayed as a *grey solid line*

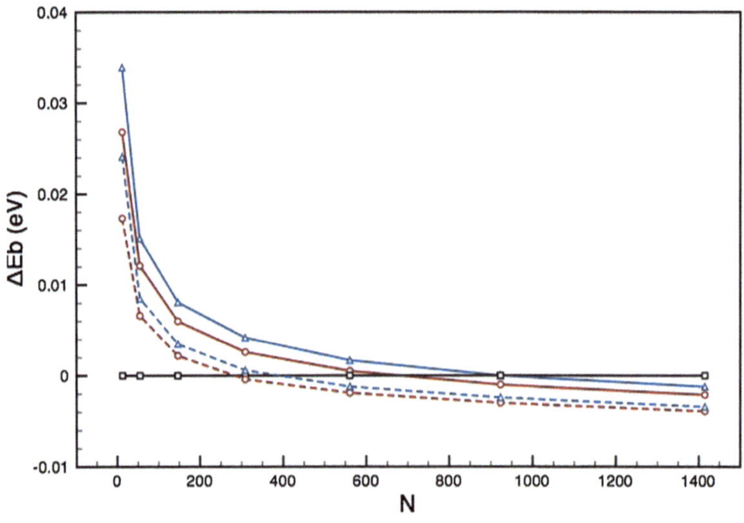

Fig. 2.11 Plot of ΔE_b for Ih (*black*), I-Dh (*red*) and CO (*blue*) Au_N structures against N. *Solid lines* represent parameter set I of Cleri and Rosato [12], and *dashed lines* represent parameter set II of Baletto et al. [2]

these intersections of $E_{b_{Ih}}$ are at lower N: 284 and 393 for $\Delta E_{b_{I-Dh}}$ and $\Delta E_{b_{CO}}$, respectively. The values of N for these intersections are much less than found for Pd_N, implying relative instability of the close-packed cluster geometries (Ih, I-Dh) with respect to the crystalline bulk fragments for Au nanoparticles.

FCC Structures

Comparison of the CO structures with a variety of other FCC fragments (Fig. 2.12) leads to the observation that the CO structure is not the optimal for the FCC fragments, identical to findings for Pd. Whilst a large spread of energetic values have been identified ($0.8 < \Delta < 2$ eV), some higher in energy than the CO, the energetic minima and associated trend lines are approximately 0.1 eV lower in energy than for the CO structures. Of particular interest is the positive gradient seen for the energy trend line of parameter set I, with values reaching a minimum at $\Delta = 1.269$ eV for $N = 79$ ($n_l = 5, n_{cut} = 1$), before rising towards 1.5 eV. This case is not the same for the parameter set II, where a plateau is reached for $N > 1{,}000$ at $\Delta \sim 0.95$ eV. Quantitatively, the Δ values of parameter set I are greater than those of parameter set II. The minimum atom-centred and void-centred geometries are closely competitive in energy for both parameter sets; further analysis shows that the preferred structures maximise the hexagonal (111)-packed surfaces, whilst minimising exposed (100) surfaces. This observation can be linked to the surface energies of the (111) and (100) surfaces for Au, which have previously been reported by Uppenbrink et al. as 93.2 and 97.1 meV Å^{-2}, respectively [56], thus favouring Au (111) facet formation over (100).

Marks-Decahedra

M-Dh structures have been generated, using systematic variation of m, n and r, and minimised using both parameter sets; Δ has been plotted in Fig. 2.13 for the series of structures Au_N. Similar to the case of Pd_N, we find that fully stellated M-Dh structures are unfavourable (except inevitably for Au_{18} when $m = n = r = 1$, as truncation is not possible). The formation of truncated features is favourable compared to the simple I-Dh clusters, but the ratio $\frac{r}{r_{max}}$ tends to be smaller than previously seen for Pd_N at large N, indicating that large stellated growths on the (100) faces, truncated or not, are generally not energetically suitable; e.g. for $r_{max} = 9$ a truncation of the edges such that $m = n = 7$, $r = 3$. The trend lines for each value of r_{max} are similar to Pd_N, i.e. they are steeper sided funnels at low N, which tend to flatten out as $N \rightarrow 6{,}000$. The results values using parameter set I rise above those of parameter set II, a reverse of the trend for Pd_N M-Dh structures in Fig. 2.7, and similar to our observations noted for Figs. 2.10, 2.11 and 2.12 in relation to the energetic ordering of parameter sets I and II. We also note here that Δ increases with N for parameter set I, similar to the top of Fig. 2.12.

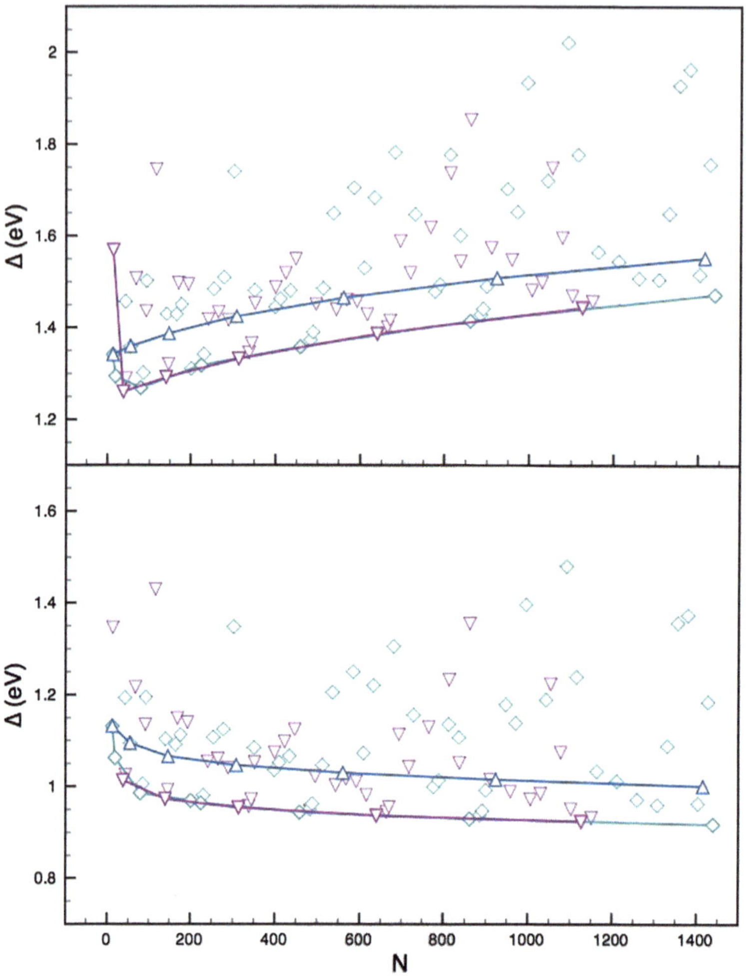

Fig. 2.12 *Top*: Plot of Δ against N for various TO Au_N structures, where $N < 1{,}500$, for parameter set I of Cleri and Rosato [12]. *Bottom*: As above, using parameter set II of Baletto et al. [2]. FCC structures with an atom at their centre (*green diamonds*) and with a void at their centre (*purple triangles*) are plotted against the CO structure (*blue upwards triangles*) of increasing shell size k. *Lines* are plotted to show the lowest values for each structure type. Reproduced from [55] by permission of the PCCP Owner Societies

Au_{887}

A selection of the most energetically favourable Δ for Ih, I-Dh, CO, TO and M-Dh are plotted against N in Fig. 2.14, and compared to Au_{887}. Similar to Pd_{887}, Au_{887} has been the subject of recent experimental interest [35]. For both parameter sets I and II we find that Δ for the 12-vertex high-symmetry structures (Ih, I-Dh and CO)

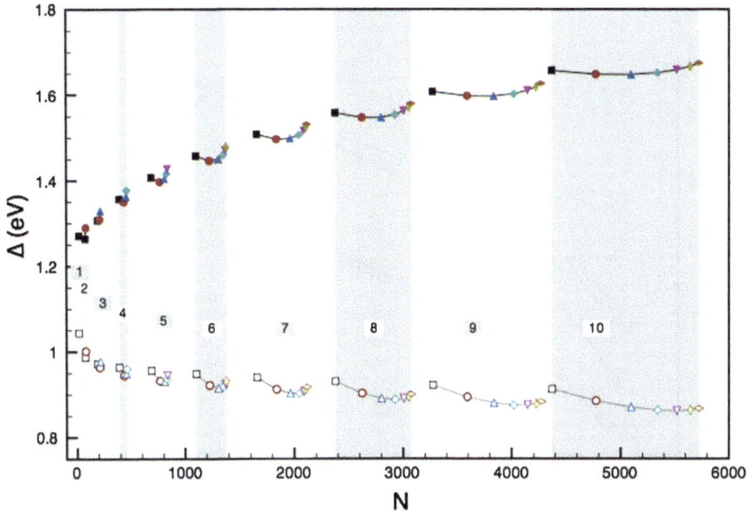

Fig. 2.13 Plot of Δ against N for Au_N M-Dh of varying re-entrant sizes. Structures have been created such that $r < 9$, and $m = n < 11$ where $r = 2$ (*black squares*), 3 (*red circles*), 4 (*blue upward triangles*), 5 (*green diamonds*), 6 (*purple downward triangles*), 7 (*yellow laterally-extended diamonds*) and 8 (*brown vertically-extended diamonds*) are displayed. Parameter set I, of Cleri and Rosato [12], is plotted with *closed symbols* and *black trend lines*, and parameter set II, of Baletto et al. [2], is plotted with *open symbols* and *grey trend lines*. *Trend lines* illustrate the effect of reduced truncation of the stellated features from an I-Dh core ($r = 1$) to fully stellated structures ($m = 1$). Background colouring is added to highlight distinct trend lines. r_{max} is also labelled from 1 to 10. Reproduced from [55] by permission of the PCCP Owner Societies

is 0.1 eV greater than the most stable FCC and M-Dh structures, and that Au_{887} is energetically closer to the latter. The Ih structure is closely competitive for $N < 100$, before becoming strongly unfavourable for $N > 500$. Δ for the other investigated structures run parallel to each other for $N > 200$ for both parameter sets; below this they are closer in energy, though one would also suspect that there are many other alternative structures which we have not investigated here which may be competitive. The heat required for direct thermal transformation from Ih/I-Dh/CO to M-Dh/TO is again in the range of $T \approx 1{,}200$ K. This value is sufficiently high that we would not expect structural rearrangement in ambient experimental conditions; for smaller ($N < 100$) this is not the case and structural evolution under an electron beam has been documented [32].

2.4.3 $(Au_{core}Pd_{shell})_N$ Clusters

With the 12-vertex high-symmetry structures it is possible to conduct systematic studies of their core/shell bimetallic structures by building up clusters with layers

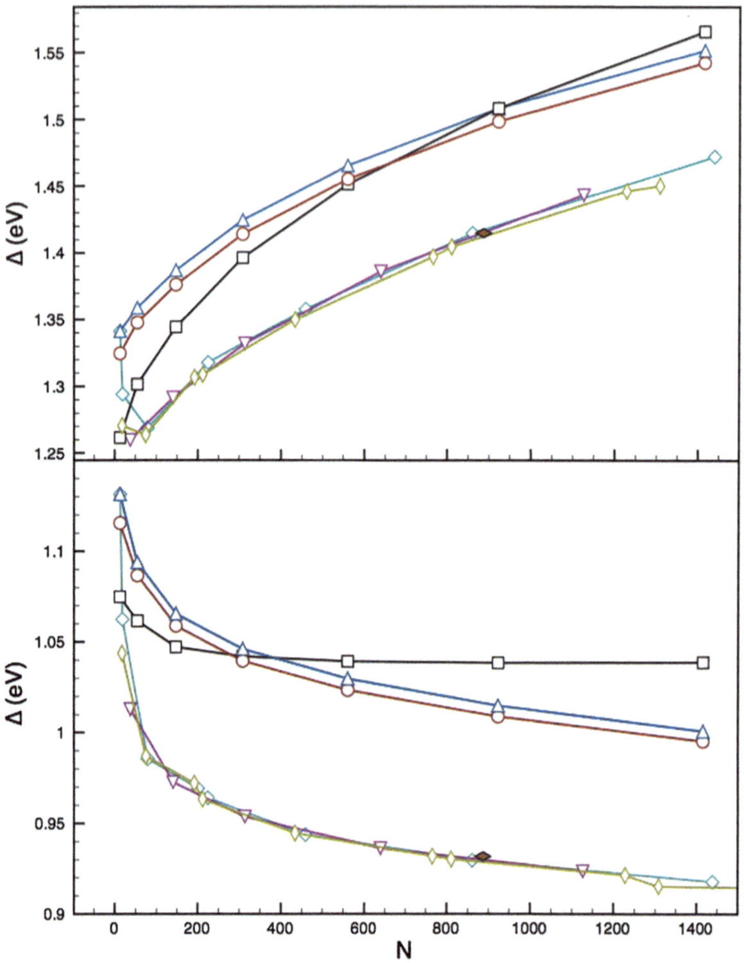

Fig. 2.14 *Top*: Plot of Δ for Au_N structures against N, using parameter set I of Cleri and Rosato [12]. *Bottom*: As above, using parameter set II of Baletto et al. [2]. Ih (*black squares*), I-Dh (*red circles*), CO (*blue upward triangles*) are given, as well as the most energetically competitive atom-centred and void-centred TO (*purple downward triangles* and *green diamonds*, respectively), and M-Dh structures (*yellow vertical diamonds*). Au_{887} is also shown (*brown filled lateral diamond, black edges*). Adapted from [55] by permission ofthe PCCP Owner Societies

of different atom types. One of the more fundamental cluster structures is that of core/shell segregation, whereby one element occupies sites close to the core of the cluster and another element occupies more surface sites. This kind of arrangement is commonly documented experimentally for $(AuPd)_N$ [3, 57]. By creating a fixed core of one element and nuclearity, and growing shells of another element on the surface (Fig. 2.15), we are able to compare the relative stability of the core/shell bimetallic systems with respect to the homogeneous monometallic systems, and also

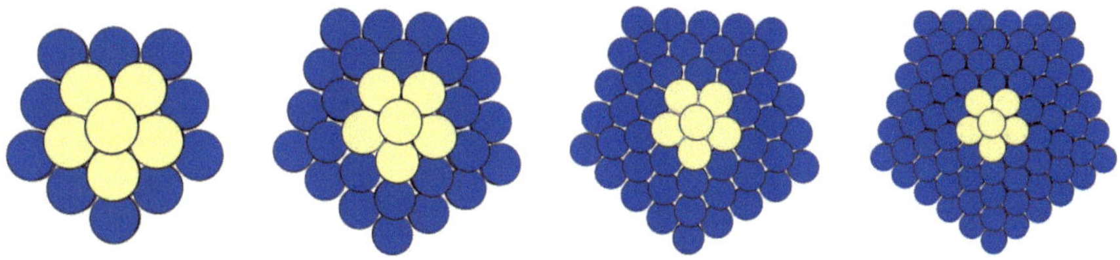

Fig. 2.15 Cross section of $(Au_{core}Pd_{shell})_N$ Dh with a fixed core of 13 atoms ($k = 1$). *Left* to *right*: N = 55, 147, 309 and 561, respectively. Pd and Au are represented in *blue* and *yellow*, respectively. Reproduced from [46] by permission of The Royal Society of Chemistry

the relative stability of the different core/shell structures to each other, i.e. the Ih, I-Dh and CO. Naively, one could expect a linear transition between the structural preferences of Au and Pd as we are using arithmetic means for the Au-Pd interaction.

Comparison of Δ for the pure clusters offers initial insight into the predicted energetic evolution with the addition of a second element to the cluster: Fig. 2.16 gives Δ for the Ih, I-Dh and CO values for Au_N and Pd_N clusters. For parameter set *I* we notice that for $N < 150$ the Au_N structures have a lower Δ than Pd_N, however this does not hold true for $N > 200$, with Pd_N becoming energetically preferred and the gap between the trend lines increasing up to $N = 1{,}500$. For parameter set *II* the Au_N structures remain more energetically stable compared to those of Pd_N, with the gap between the two remaining constant at ~0.5 eV. As discussed previously, the Ih structure is favoured to large *N* for Pd_N, unlike Au_N, and this is shown for both parameter sets.

Ih, I-Dh and CO clusters were generated with fixed-size Au cores and Pd layering added to form the shell, as illustrated in Fig. 2.15. Results from parameter set *I* and *II* have been plotted in Fig. 2.17 and 2.18, respectively. In Fig. 2.17 comparison with Pd_N shows that adding the Au impurity results in an increase in E_{exc}, and thus Δ, for all clusters; and the greater the concentration of Au in the core of the structure, the greater this difference is between Δ for Pd_N and $(Au_{core}Pd_{shell})_N$. Though not plotted, comparison with Au_N also shows that Δ for $(Au_{core}Pd_{shell})_N$ is increased with respect to Au_N when the number of Pd layers is small ($k < 4$) , making this form of mixing not preferable for the bimetallic system at low nuclearities. Larger clusters appear more stable, with Δ for the bimetallic cluster falling between the pure clusters, and the difference in Δ between Pd_N and $(Au_{core}Pd_{shell})_N$ decreasing as *N* increases. We also note that for a core of Au_{55} or greater the Ih is no longer the preferred structure with respect to the I-Dh and CO; this is in-keeping with the observed trends for the pure Au_N clusters.

For parameter set *II* (Fig. 2.18), this destabilised effect is not so prominent. Whilst thin (monolayer) Pd shells are not energetically favourable, especially for increasing size of the Au_N core, multiple Pd layers have a quenching effect on this disparity, reducing E_{exc} such that for an Au_{13} core with three layers of Pd or greater ($k > 3$) it appears more energetically favourable than the Pd_N clusters, though not also the Au_N clusters.

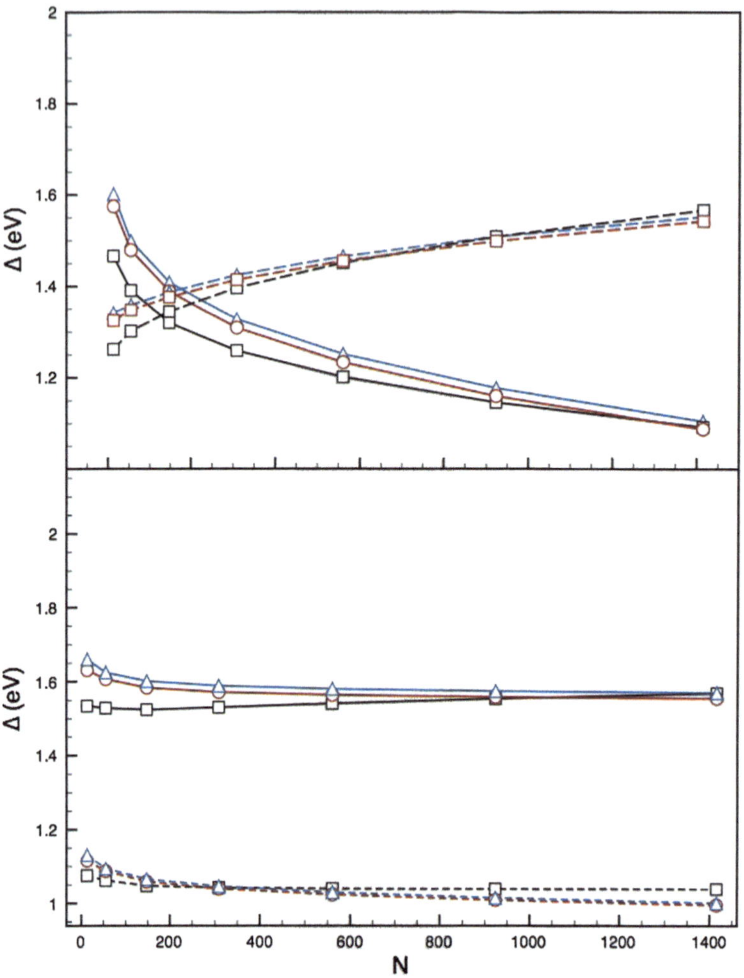

Fig. 2.16 *Top*: Plot of Δ against *N* for the 12-vertex high-symmetry structures, using parameter set *I* of Cleri and Rosato [12]. *Bottom*: As above, using parameter set *II* of Baletto et al. [2]. Ih (*black squares*), I-Dh (*red circles*) and CO (*blue triangles*) are shown, with Pd_N (*solid lines*) and Au_N (*dashed lines*) plotted. Reproduced from [46] by permission of The Royal Society of Chemistry

2.4.4 $(Pd_{core}Au_{shell})_N$ Clusters

Using parameter set *I* and inverting the core-shell segregation to form $(Pd_{core}Au_{shell})_N$ leads to the reverse effect of that witnessed for the $(Au_{core}Pd_{shell})_N$: the value of Δ tends to decrease with respect to the pure clusters (Fig. 2.19) as the size of the Pd core increases. The effect is greatest for smaller clusters, with monolayers of Au, with the difference between Δ of the core/shell and pure clusters reducing as *N* increases, for a fixed core. For multiple Au shell layers Δ shifts to a value

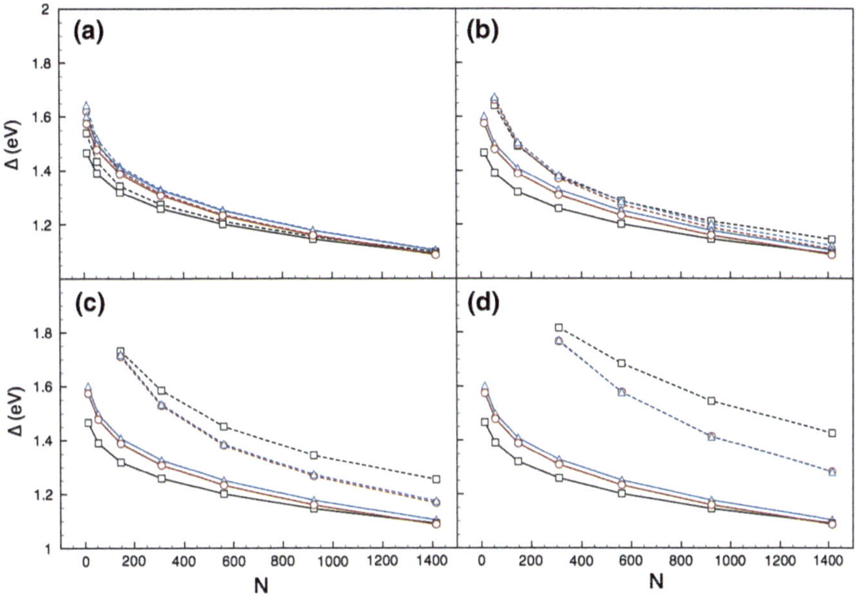

Fig. 2.17 Plots of Δ against N for the $Au_{core}Pd_{shell}$ 12-vertex high-symmetry structures (*dashed lines*), using parameter set I of Cleri and Rosato [12], with cores: **a** Au_1 **b** Au_{13} **c** Au_{55} **d** Au_{147}. Ih (*black squares*), I-Dh (*red circles*) and CO (*blue triangles*) are shown, with Pd_N (*solid lines*) also plotted. Reproduced from [46] by permissionof The Royal Society of Chemistry

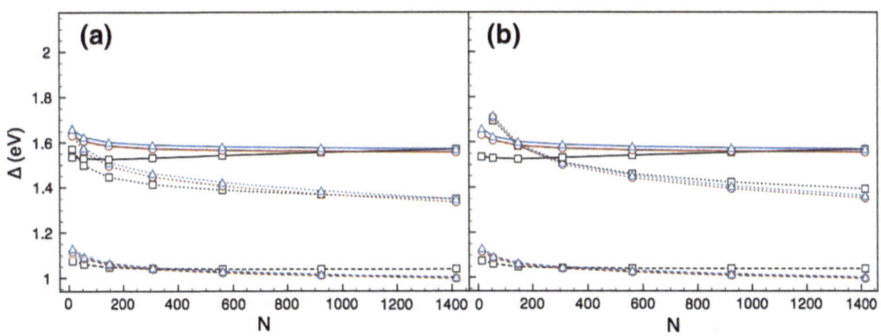

Fig. 2.18 Plots of Δ against N for the $Au_{core}Pd_{shell}$ 12-vertex high-symmetry structures (*dotted lines*), using parameter set II of Baletto et al. [2], with cores: **a** Au_1 and **b** Au_{13}. Ih (*black squares*), I-Dh (*red circles*) and CO (*blue triangles*) are shown, with Pd_N (*solid lines*) and Au_N (*dashed lines*) also plotted. Reproduced from [46] by permission of The Royal Society of Chemistry

between the pure values, illustrating again the averaging effect we spoke of previously. Figure 2.19 also shows that the Ih structure becomes distinctly more preferable than the I-Dh and CO structures, whilst the energy gap between the I-Dh and CO also increases. For a core of Pd_{55} or greater the transition from Ih to I-Dh and CO is out of the nuclearity range studied (i.e. $N > 1{,}500$).

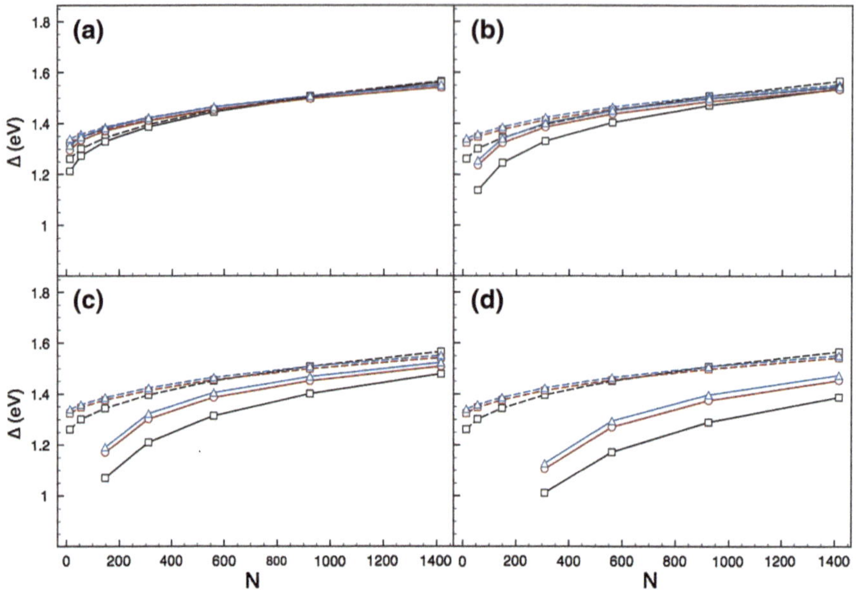

Fig. 2.19 Plots of Δ against N for the $Pd_{core}Au_{shell}$ 12-vertex high-symmetry structures (*solid lines*), using parameter set I of Cleri and Rosato [12], with cores: **a** Pd_1 **b** Pd_{13} **c** Pd_{55} **d** Pd_{147}. Ih (*black squares*), I-Dh (*red circles*) and CO (*blue triangles*) are shown, with Au_N (*dashed lines*) also plotted

For parameter set II the effect of multiple Au layers is not so favourable (Fig. 2.20). Small Pd impurities in the core of small Au clusters are energetically favourable, compared to the pure Pd_N clusters, but not compared to the Au_N clusters. For all core sizes, an Au coating becomes unfavourable with respect to the pure clusters when thicker than a monolayer, and increases with thicker Au coverage, illustrating that the segregation we have used may not be as preferable as a mixed alloy with these parameters, similar to results seen by Ismail et al. [39].

Attention to the relative stability of the $(Pd_{core}Au_{shell})_N$ structures with just a monolayer coating, as shown in Fig. 2.21, shows an unusually high stability which increases with N (Fig. 2.22) for both parameter sets I and II.

A monolayer coating of Au on a Pd core, with parameter set I, results in Δ less than for both Au_N and Pd_N clusters of the same nuclearity. This is a strong indication of the preference of the stability of the $Pd_{core}Au_{shell}$ segregation, and agrees with the common discovery of core-shell structures in previous structural searches [38, 39]. The difference between the energy of the Pd_N and the $(Pd_{core}Au_{shell})_N$ remains constant as k increases, at a gap of $\sim$0.3 eV. The Ih structure is energetically preferable throughout, in stark contrast to the same calculations for the Au_N clusters which show an early tendency to adopt crystalline structures.

For parameter set II a monolayer coating has a similar effect in stabilising the Pd_N cluster. However, for $N < 800$ ($k < 6$) the $(Pd_{core}Au_{shell})_N$ clusters are less stable

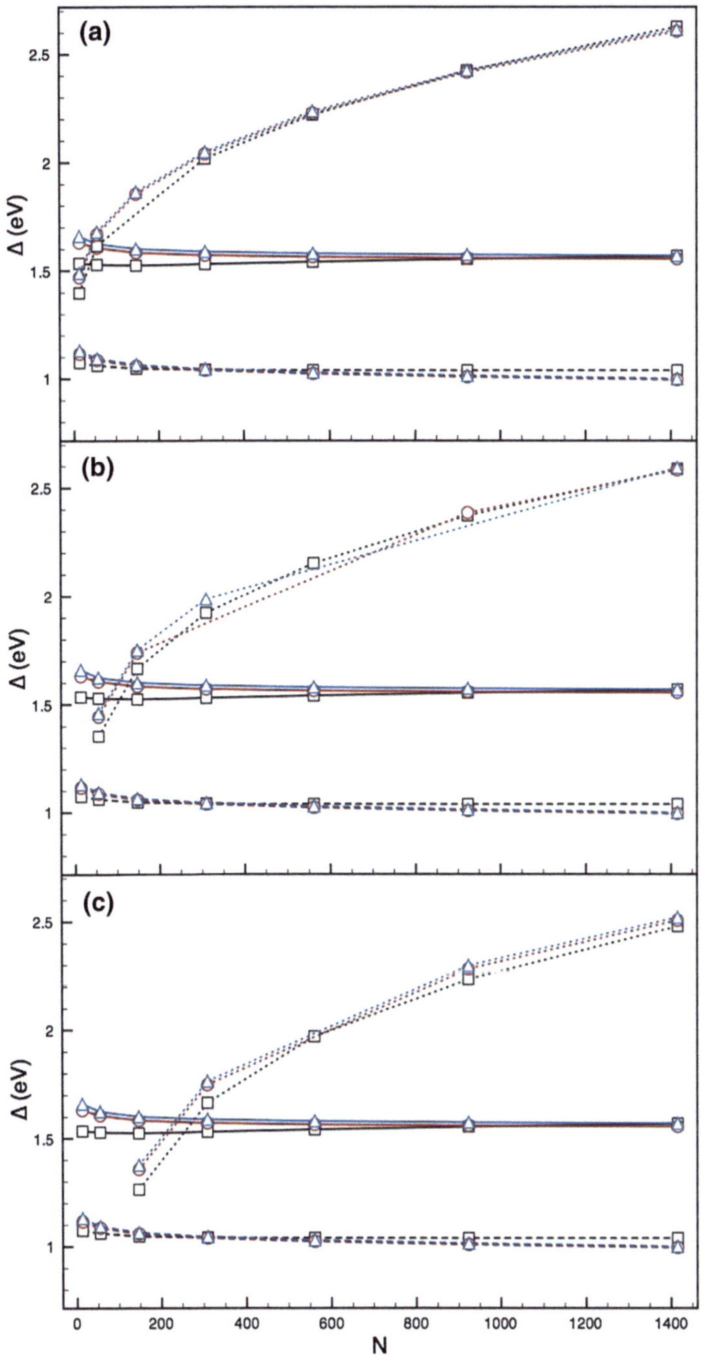

Fig. 2.20 Plots of Δ against N for the $Pd_{core}Au_{shell}$ 12-vertex high-symmetry structures (*dotted lines*), using parameter set *II* of Baletto et al. [2], with cores: **a** Pd_1 **b** Pd_{13} and **c** Pd_{55}. Ih (*black squares*), I-Dh (*red circles*) and CO (*blue triangles*) are shown, with Au_N (*dashed lines*) and Pd_N (*solid lines*) also plotted. Reproduced from [46] by permissionof The Royal Society of Chemistry

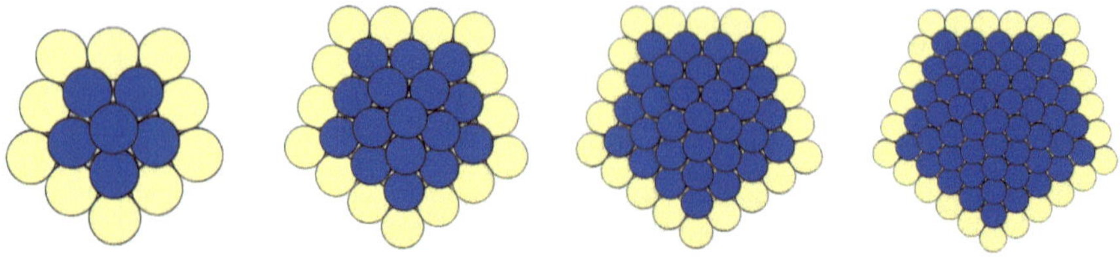

Fig. 2.21 Cross section of $(\text{Pd}_{core}\text{Au}_{shell})_N$ Dh with a monolayer covering on the shell. *Left* to *right*: N = 55, 147, 309 and 561, respectively. Pd and Au are represented in *blue* and *gold*, respectively. Reproduced from [46] by permission of The Royal Society of Chemistry

than the pure Au_N clusters. For $N > 800$ the core/shell segregation becomes more energetically favourable than the pure Au_N cluster; this implies that this segregation is not-preferable at low N, where mixing is perhaps preferred, but as N increases, and so the ratio of Pd to Au atoms increases in favour of Pd, so does the suitability of core/shell segregation to this structure. This is perhaps the driving force behind the structural rearrangements seen in previous molecular dynamics studies [41], where core/shell inversion occurs. As the difference in Δ between the pure clusters and $(\text{Pd}_{core}\text{Au}_{shell})_N$ is increasing, we can predict that as N increases beyond $N = 1{,}500$ that this structural arrangement may become even more favourable, as it minimises E_{exc}.

2.5 Conclusions

We have studied the stabilities of high-symmetry Au_N, Pd_N and $(\text{AuPd})_N$ clusters, using mathematical constructs, a semi-empirical potential with two different parameter sets, and a quasi-newtonian minimisation technique. The structures for which calculations were performed include the Ih, I-Dh, CO, M-Dh and FCC fragments; this list is by no means exhaustive but does give a good insight to the preferred nanocluster packing in high-symmetry structures.

For Pd_N clusters both parameter sets tested result in preferences for the Ih structures for $N < 1{,}000$ over other high-symmetry 12-vertex geometries. However, TO Pd_N clusters with reduced (100) exposure, and M-Dh Pd_N clusters with truncation stellations prove to have lower E_{exc}, as a result of the reduced (100) surfaces. Pd_{887} proves to be competitive with M-Dh and TO structures when using parameter set II; this is not the case for parameter set I where Pd_{887} is similar in Δ to the 12-vertex structures. Overall, the parameters of Cleri and Rosato tend towards values of E_b greater than the parameters of Baletto et al., with extrapolation to the bulk limit illustrating the latter gives results closer to the experimentally measured value.

For Au_N clusters we found a tendency towards FCC structures at values of N lower than seen for Pd_N: the parameters of Cleri and Rosato gave a transition at $N \approx 650$ to the I-Dh, whilst for the parameters of Baletto et al. this value was lower

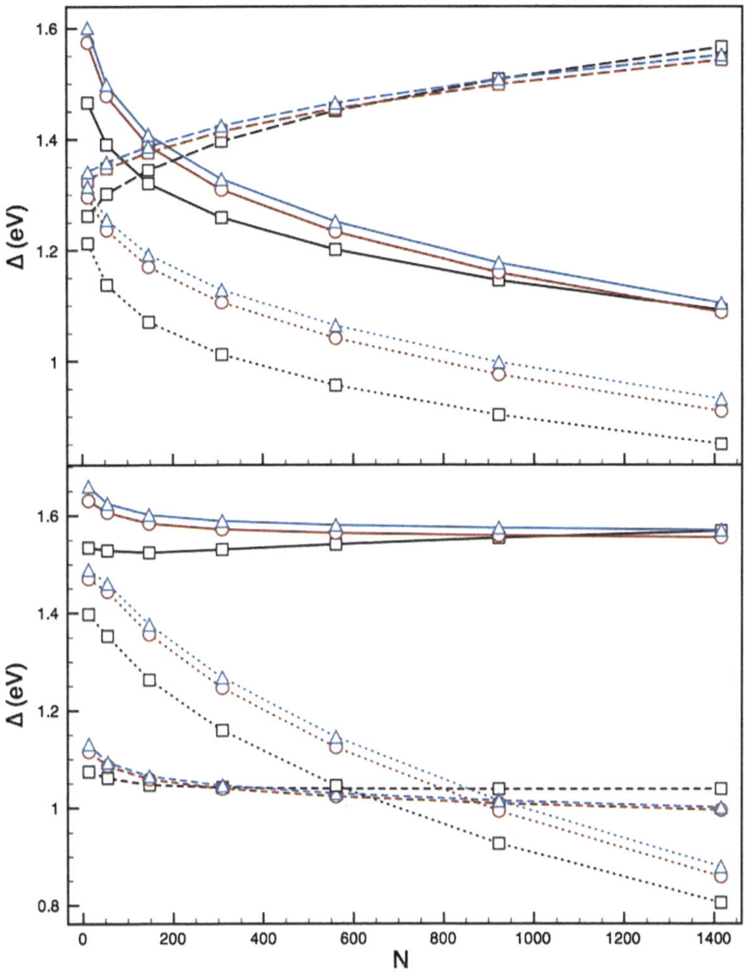

Fig. 2.22 *Top*: Plots of Δ against *N* for the $Au_{core}Pd_{shell}$ 12-vertex high-symmetry structures, using parameter set *I* of Cleri and Rosato [12], with a monolayer covering of Au. *Bottom*: As above, using parameter set *II* of Baletto et al. [57]. Ih (*black squares*), I-Dh (*red circles*) and CO (*blue triangles*) are shown; Pd_N (*solid lines*), Au_N (*dashed lines*) and $Pd_{core}Au_{shell}$ (*dotted lines*) are plotted. Reproduced from [46] by permission of The Royal Society of Chemistry

still. TO and M-Dh structures again prove more stable than the 12-vertex geometries, with shrinking of the (100) faces favourable. Au_{887} is competitive with the energetic minima for both parameter sets, offering enthusiasm for further experimental work trying to identify this structure. Overall, the parameters of Baletto et al. give a larger value of E_b at low *N* than the parameters of Cleri and Rosato.

For $(AuPd)_N$ clusters we found that the preferred arrangement is for $(Pd_{core}Au_{shell})_N$ for both parameter sets, with thin (monolayer) surface coverings of Au being most energetically favourable compared to the homogeneous clusters; this is highly compatible with the experimental measurements $E_{surf}(\text{Pd}) > E_{surf}(\text{Au})$ and $E_{coh}(\text{Pd}) > E_{coh}(\text{Au})$. For the parameters of Baletto et al. multiple layers of Au lead to energetic instability, with Δ values greater than for either of the pure metal nanoparticles; the same is not the case for the parameters of Cleri and Rosato. $(\text{Au}_{core}\text{Pd}_{shell})_N$ clusters are not energetically favourable with thin coatings of Pd, however as the shell coating thickens so the stability improves. This effect is more prominent for the parameters of Cleri and Rosato. Ih structures are not favourable compared to the I-Dh and CO for $(\text{Au}_{core}\text{Pd}_{shell})_N$, where as they are strongly preferred for $(\text{Pd}_{core}\text{Au}_{shell})_N$. Overall, the strong tendency towards core/shell segregation is emphasised for the parameters of Cleri and Rosato [12], and not so much for the parameters of Baletto et al. [2], agreeing with previous observations for smaller $(\text{AuPd})_N$ clusters [38, 39].

Exploring other high-symmetry homotops for the bimetallic $(\text{AuPd})_N$ clusters is necessary in the future to know whether other segregated formations are more favourable than those postulated here, with investigation into other lower-symmetry structures also required to give a more complete treatment of structural preference in any future work.

References

1. R.L. Johnston, *Atomic and Molecular Clusters* (Taylor and Francis, London, 2002)
2. F. Baletto, R. Ferrando, A. Fortunelli, F. Montalenti, C. Mottet, Crossover among structural motifs in transition and noble-metal clusters. J. Chem. Phys. **116**(9), 3856–3863 (2002)
3. R. Ferrando, J. Jellinek, R.L. Johnston, Nanoalloys: from theory to applications of alloy clusters and nanoparticles. Chem. Rev. **108**(3), 845–910 (2008)
4. F. Baletto, R. Ferrando, Structural properties of nanoclusters: energetic, thermodynamic, and kinetic effects. Rev. Mod. Phys. **77**, 371–423 (2005)
5. P. Nava, M. Sierka, R. Ahlrichs, Density functional study of palladium clusters. Phys. Chem. Chem. Phys. **5**, 3372–3381 (2003)
6. G. Valerio, H. Toulhoat, Atomic sulfur and chlorine interaction with Pd_n clusters (n = 1–6): a density functional study. J. Phys. Chem. A **101**(10), 1969–1974 (1997)
7. B. Kalita, R.C. Deka, Stability of small Pd_n (n = 1–7) clusters on the basis of structural and electronic properties: a density functional approach. J. Chem. Phys. **127**(24), 244306–244316 (2007)
8. D.R. Jennison, P.A. Schultz, M.P. Sears, Ab initio calculations of Ru, Pd, and Ag cluster structure with 55, 135, and 140 atoms. J. Chem. Phys. **106**(5), 1856–1862 (1997)
9. R. Guirado-López, M.C. Desjonquères, D. Spanjaard, Tight-binding study of relaxation in Rh_N and Pd_N clusters ($9 < N < 165$). Phys. Rev. B **62**(19), 13188–13195 (2000)
10. F. Aguilera-Granja, A. Vega, J. Rogan, G. Garcia, Metallic behavior of Pd atomic clusters. Nanotechnology **18**(36), 365706–365711 (2007)
11. N. Watari, S. Ohnishi, Atomic and electronic structures of Pd_{13} and Pt_{13} clusters. Phys. Rev. B **58**(3), 1665–1677 (1998)
12. F. Cleri, V. Rosato, Tight-binding potentials for transition metals and alloys. Phys. Rev. B **48**(1), 22–33 (1993)
13. V. Rosato, M. Guillope, B. Legrand, Thermodynamical and structural properties of FCC transition metals using a simple tight-binding model. Philos. Mag. A **59**(2), 321–336 (1989)

14. M. José-Yacamán, M. Marin-Almazo, J.A. Ascencio, High resolution TEM studies on palladium nanoparticles. J. Mol. Catal. A **173**(1–2), 61–74 (2001)
15. M. José-Yacamán, J.A. Ascencio, H.B. Liu, J. Gardea-Torresdey, Structure shape and stability of nanometric sized particles. J. Vac. Sci. Technol. B **19**(4), 1091–1103 (2001)
16. S.M. Morton, D.W. Silverstein, L. Jensen, Theoretical studies of plasmonics using electronic structure methods. Chem. Rev. **111**, 3962–3994 (2011)
17. G. Bravo-Perez, I.L. Garzan, O. Novaro, Non-additive effects in small gold clusters. Chem. Phys. Lett. **313**(3–4), 655–664 (1999)
18. J. Wang, G. Wang, J. Zhao, Density-functional study of Au_n (n=2–20) clusters: lowest-energy structures and electronic properties. Phys. Rev. B **66**(3), 35418–35424 (2002)
19. V. Bonačić-Koutecký, J. Burda, R. Mitrić, M. Ge, G. Zampella, P. Fantucci, Density functional study of structural and electronic properties of bimetallic silver-gold clusters: comparison with pure gold and silver clusters. J. Chem. Phys. **117**(7), 3120–3131 (2002)
20. J. Li, X. Li, H.J. Zhai, L.S. Wang, Au_{20}: a tetrahedral cluster. Science **299**(5608), 864–867 (2003)
21. J. Wang, G. Wang, J. Zhao, Structures and electronic properties of Cu_{20}, Ag_{20}, and Au_{20} clusters with density functional method. Chem. Phys. Lett. **380**, 716–720 (2003)
22. K. Michaelian, N. Rendón, I.L. Garzón, Structure and energetics of Ni, Ag, and Au nanoclusters. Phys. Rev. B **60**, 2000–2010 (1999)
23. C.L. Cleveland, U. Landman, M.N. Shafigullin, P.W. Stephens, R.L. Whetten, Structural evolution of larger gold clusters. Z. Phys. D **40**, 503–508 (1997)
24. N.T. Wilson, R.L. Johnston, Modelling gold clusters with an empirical many-body potential. Eur. Phys. J. D. **12**(1), 161–169 (2000)
25. M.M. Alvarez, J.T. Khoury, T.G. Schaaff, M.N. Shafigullin, I. Vezmar, R.L. Whetten, Optical absorption spectra of nanocrystal gold molecules. J. Phys. Chem. B **101**(19), 3706–3712 (1997)
26. H. Häkkinen, R.N. Barnett, U. Landman, Electronic structure of passivated $Au_{38}(SCH3)_{24}$ nanocrystal. Phys. Rev. Lett. **82**(16), 3264–3267 (1999)
27. H. Häkkinen, M. Walter, H. Grönbeck, Divide and protect: capping gold nanoclusters with molecular gold-thiolate rings. J. Phys. Chem. B **110**(20), 9927–9931 (2006)
28. J. Akola, M. Walter, R.L. Whetten, H. Häkkinen, H. Grönbeck, On the structure of thiolate-protected Au_{25}. J. Am. Chem. Soc. **130**(12), 3756–3757 (1999)
29. M. Walter, J. Akola, O. Lopez-Acevedo, P.D. Jadzinsky, G. Calero, C.J. Ackerson, R.L. Whetten, H. Grönbeck, H. Häkkinen, A unified view of ligand-protected gold clusters as superatom complexes. Proc. Natl. Acad. Sci. **105**, 9157–9162 (2008)
30. O. Lopez-Acevedo, J. Akola, R.L. Whetten, H. Grönbeck, H. Häkkinen, Structure and bonding in the ubiquitous icosahedral metallic gold cluster $Au_{144}(SR)_{60}$. J. Phys. Chem. C **113**, 5035–5038 (2009)
31. O. Lopez-Acevedo, H. Tsunoyama, T. Tsukuda, H. Häkkinen, C.M. Aikens, Chirality and electronic structure of the thiolate-protected Au_{38} nanocluster. J. Am. Chem. Soc. **132**(23), 8210–8218 (2010)
32. Z. Wang, O. Toikkanen, B.M. Quinn, R.E. Palmer, Real-space observation of prolate monolayer-protected Au_{38} clusters using aberration-corrected scanning transmission electron microscopy. Small **7**(11), 1542–1545 (2011)
33. Z.Y. Li, N.P. Young, M. Di Vece, S. Palomba, R.E. Palmer, A.L. Bleloch, B.C. Curley, R.L. Johnston, J. Jiang, J. Yuan, Three-dimensional atomic-scale structure of size-selected gold nanoclusters. Nature **451**(7174), 46–48 (2008)
34. B.C. Curley, R.L. Johnston, N.P. Young, Z.Y. Li, M. Di Vece, R.E. Palmer, A.L. Bleloch, Combining theory and experiment to characterize the atomic structures of surface-deposited Au_{309} clusters. J. Phys. Chem. C **111**(48), 17846–17851 (2007)
35. N.J. Cookson, *Preparation and Characterisation of Bimetallic Core-Shell Nanoparticles*, School of Chemistry, University of Birmingham, Edgbaston, Birmingham, Master's thesis, 2009
36. M.J. Cabrera-Trujillo, J.M. Montejano-Carrizales, J.L. Rodriguez-Lopez, W. Zhang, J.J. Velazquez-Salazar, M. José-Yacamán, Nucleation and growth of stellated gold clusters: experimental synthesis and theoretical study. J. Phys. Chem. C **114**(49), 21051–21060 (2010)

37. R.L. Johnston, Evolving better nanoparticles: genetic algorithms for optimising cluster geometries. Dalton Trans. **22**, 4193 (2003)
38. F. Pittaway, L.O. Paz-Borbón, R.L. Johnston, H. Arslan, R. Ferrando, C. Mottet, G. Barcaro, A. Fortunelli, Theoretical studies of palladium-gold nanoclusters: Pd-Au clusters with up to 50 atoms. J. Phys. Chem. C **113**, 9141–9152 (2009)
39. R. Ismail, R.L. Johnston, Investigation of the structures and chemical ordering of small pd-au clusters as a function of composition and potential parameterisation. Phys. Chem. Chem. Phys. **12**(30), 8607–8619 (2010)
40. B. Shan, L. Wang, S. Yang, J. Hyun, N. Kapur, Y. Zhao, J.B. Nicholas, K. Cho, First-principles-based embedded atom method for PdAu nanoparticles. Phys. Rev. B **80**, 035404 (2009)
41. H.B. Liu, U. Pal, A. Medina, C. Maldonado, J.A. Ascencio, Structural incoherency and structure reversal in bimetallic Au–Pd nanoclusters. Phys. Rev. B **71**, 075403 (2005)
42. A.R. Miedema, Surface energy of solid metals. Z. Metal. **69**(5), 287–292 (1978)
43. D. Ferrer, A. Torres-Castro, X. Gao, S. Sepúlveda-Guzmán, U. Ortiz-Méndez, M. José-Yacamán, Three-layer core/shell structure in Au-Pd bimetallic nanoparticles. Nano Lett. **7**, 1701–1705 (2007)
44. T.P. Martin, Shells of atoms. Phys. Rep. **273**, 199–241 (1996)
45. X. Xing, B. Yoon, U. Landman, J.H. Parks, Structural evolution of Au nanoclusters: From planar to cage to tubular motifs. Phys. Rev. B **74**(16), 165423 (2006)
46. A.J. Logsdail, R.L. Johnston, Interdependence of structure and chemical order in high symmetry $(\mathrm{pdau})_N$ nanoclusters. RSC Adv. **2**, 5863–5869 (2012)
47. J.N. Murrell, R.E. Mottram, Potential energy functions for atomic solids. Mol. Phys. **69**, 571–585 (1990)
48. A.P. Sutton, J. Chen, Long-range Finnis-Sinclair potentials. Philos. Mag. Lett. **61**(3), 139–146 (1990)
49. M.S. Daw, M.I. Baskes, Embedded-atom method: derivation and application to impurities, surfaces, and other defects in metals. Phys. Rev. B **29**(12), 6443–6453 (1984)
50. R.P. Gupta, Lattice relaxation at a metal surface. Phys. Rev. B **23**(12), 6265–6270 (1981)
51. R. Ferrando. Personal communication, 2009.
52. C. Kittel, *Introduction To Solid State Physics*, 6th edn. (Wiley, New York, 1986)
53. C.L. Cleveland, U. Landman, The energetics and structure of nickel clusters: size dependence. J. Chem. Phys. **94**(11), 7376–7396 (1991)
54. J. Uppenbrink, D.J. Wales, Structure and energetics of model metal clusters. J. Chem. Phys. **96**(11), 8520–8534 (1992)
55. A.J. Logsdail, Z.Y. Li, R.L. Johnston, Faceting preferences for au_N and pd_N nanoclusters with high-symmetry motifs. Phys. Chem. Chem. Phys. **15**, 3473 (2013)
56. J. Uppenbrink, R.L. Johnston, J.N. Murrell, Modelling transition metal surfaces with empirical potentials. Surf. Sci. **304**(1–2), 223–236 (1994)
57. Nanoalloys: From theory to applications. Faraday Discussions No.138, vol. 138, 2008.

Chapter 3
Method Development for Comparing Scanning Transmission Electron Microscope Images to Theoretical Structures

3.1 Introduction

Scanning transmission electron microscopy (STEM), in the mode when incoherently scattered electrons are collected by a high-angle annular dark field (HAADF) detector, is appealing as a technique for probing the 3-dimensional (3D) structural properties of nanoclusters as the image intensity is proportional to both the atomic column height [1] and atomic number (Z) [2] of the component atoms. Spherical aberration (C_s) correction has recently enabled HAADF-STEM image resolution on the atomic scale [3–5], making nanocluster structure identifiable (Fig. 3.1). The process of combining experiment and simulation to assign nanostructures is still difficult, due to the complexities in identifying the orientation of nanoclusters, as well as limitations in experimental resolutions, and novel solutions continue to be sought [1, 4, 6–9].

HAADF-STEM simulations are often used quantitatively alongside experimental work: structures are generated, for example using mathematical constructs and an empirical potential, before being subjected to first-principle kinematic (Bloch Wave [6]) or dynamic [7] (such as Fast Fourier Transform [FFT] Multislice [10]) simulations. The Multislice approach has also been shown to be much more reliable for imaging non-periodic structures [11], due to its suitability for use in environments with a mobile probe, i.e. non-periodic, however both methods are computationally expensive, and coarser approximative techniques continue to be sort in order to reduce calculation time.

Li et al. used a simple, approximative, kinematic modelling technique to great success for Au_{309} clusters [1]. Firstly, idealised structures were created using the Gupta potential [12] for Icosahedral (Ih), Ino-Decahedral (I-Dh) and Cuboctahedral (CO) geometries, and were then locally relaxed. A detailed genetic algorithm search [13] was also conducted to find low-symmetry optimal structures. The resultant structures were subject to approximative kinematic, and dynamic multislice, STEM simulations. The size, 3D shape, orientation and atomic arrangement of the size-selected gold nanoclusters was determined, with atomic resolution. Wang et al. have compared images of smaller monolayer-protected Au_{38} (MP-Au_{38}) clusters

A. J. Logsdail, *Computational Characterisation of Gold Nanocluster Structures*, Springer Theses, DOI: 10.1007/978-3-319-01493-7_3,

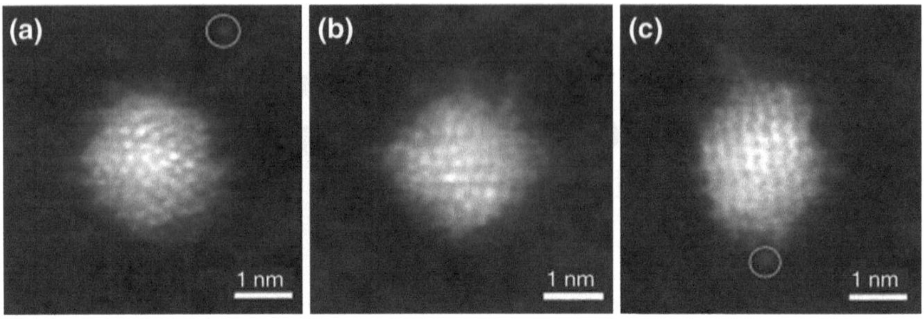

Fig. 3.1 **a** Pentagon, **b** square and **c** hexagon cluster projections identified by Li et al. for Au_{309}. The intensity variation within the clusters demonstrates atomic column resolution. *Circled* are single Au atoms close to the clusters. Adapted from Ref. [1] with permission from Macmillan Publishers Ltd (Nature), copyright 2008

to theoretical geometries using multislice, with atomic structural characterisation complicated by the mobility of clusters under the electron beam in experimental conditions [14].

The José-Yacamán group have found indications of kinks, terraces and steps on the surface of large $(PdAu)_N$ nanoparticles using first-principle HAADF-STEM simulations in conjunction with experimental work [8]. They concluded that the presence of more than one metal induces these formational defects, which may improve the catalytic properties of this system. Mkhoyan et al. have used the multislice method to show that amorphous Si layers on a crystalline Si surface distinctly impairs the clarity of STEM simulated images, with scattered electrons creating a Gaussian background and making atomic columns more difficult to see. For aberration-uncorrected probes a 20 nm amorphous layer is necessary to reduce image quality significantly, whereas for aberration-corrected probes only a 6 nm layer of amorphous Si is necessary to result in the same effect [15].

In this chapter we discuss the development of methods for identification of cluster morphologies from experimental HAADF-STEM intensity maps of substrate supported nanoclusters, in order to elucidate the possible structures. A simple kinematic model for HAADF-STEM images is presented and developed. The model is coupled with a global minimisation technique in order to minimise the computational expense in searching for matching orientation(s) of theoretically generated clusters to experimental HAADF-STEM images, using statistical comparative methods to give a measure of image similarity.

3.2 Methodology

3.2.1 STEM Model

A model has previously been proposed for simple kinematic simulation of HAADF-STEM images by Curley et al. for monometallic systems [1, 16]. For each point on a 2D grid the HAADF-STEM model intensity is calculated for a theoretical structure by summing each atom's contribution to the intensity, as determined by a Gaussian function. For a cluster defined on a Cartesian coordinate system, the contribution of each atom i to the intensity at a given point on the intensity map is given by:

$$I_{at}(i) \propto e^{-\alpha d_i^2} \tag{3.1}$$

where d_i is the distance between the current grid point and the centre of atom i on the (x, y) coordinate grid, and α is the gaussian constant which determines the decay of the electron scattering with distance (i.e. blurring). Considering the aforementioned proportional relationship between intensity and Z, and also factoring in atomic radii, means the model can be developed to accommodate systems with multiple species:

$$I_{at}^{model}(i) = Z_i^{\beta} e^{-\alpha(\frac{d_i}{r_i})^2} \tag{3.2}$$

where r_i is the radius of atom i, Z_i is the atomic number of atom i and β is the experimentally measurable scattering constant, with a limit of $\beta = 2$ for pure Rutherford scattering. By using the value $\frac{d_i}{r_i}$ we ensure that I_{at}^{model} remains dimensionless. The total intensity at a grid point j is then given by:

$$I_{tot}^{model}(j) = \sum_i I_{at}^{model}(i) \tag{3.3}$$

We note here that I_{at}^{model} is not dependent on the z-coordinate of an atom; therefore a cluster with a plane of symmetry will appear as a mirror image of the front when viewed from the back. The approach we have developed is computationally inexpensive: profiles of a cluster, examples of which are given in Fig. 3.2, can be obtained on a desktop computer in less than a second.

3.2.2 Function Evaluation

At a numerical level, experimental HAADF-STEM images are grayscale pixel maps, with each pixel having a discrete integer value: for 8-bit grayscale images the range of values is 0–255 (0 = black, 255 = white). To compare an experimental image with those produced by our HAADF-STEM model we use x- and y-dimensions matching

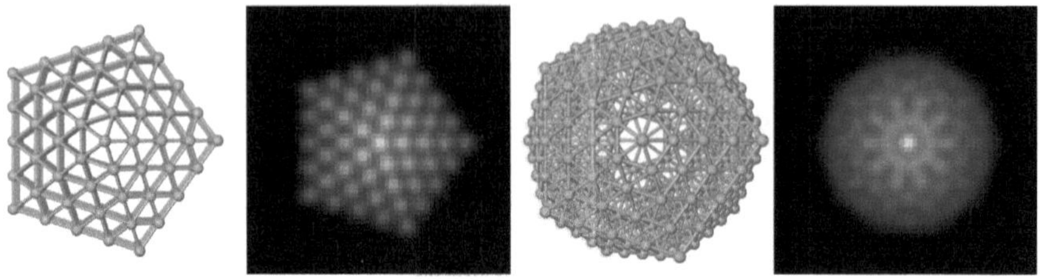

Fig. 3.2 Structures and simulated STEM images for a 309-atom cluster: From *left* to *right*: **a** an Ino-Decahedron; **b** a modelled STEM image of the same structure; **c** an Icosahedron; and **d** a modelled STEM image of the icosahedral structure. Adapted from Ref. [17] with permission from John Wiley & Sons Ltd (Journal of Computational Chemistry), copyright 2012

the experimental pixel maps; this allows the use of statistical methods for function evaluation (FE). Data comparison gives a residual (σ) with which one can compare how well modelled images match experimental images, with those returning the lowest σ values being the better results.

Covariance

The covariance allows comparison of the data sets as single entities, and can be calculated as:

$$\sigma = \overline{xy} - \bar{x} \cdot \bar{y} \tag{3.4}$$

where $\overline{xy}$, $\bar{x}$ and $\bar{y}$ are defined:

$$\bar{x} = \frac{1}{j} \sum_{j} I_{tot}^{model}(j) \tag{3.5}$$

$$\bar{y} = \frac{1}{j} \sum_{j} I_{tot}^{expt}(j) \tag{3.6}$$

$$\overline{xy} = \frac{1}{j} \sum_{j} \left[I_{tot}^{model}(j) \cdot I_{tot}^{expt}(j) \right] \tag{3.7}$$

Here I_{tot}^{expt} is the intensity map for the experimental HAADF-STEM image, and the summation is over j; the number of pixels on the intensity map.

Least Squares Fitting

Least squares fitting (LSF) compares positionally alike (x, y) points within a data set, giving an overall σ value for the similarity of the two data sets:

$$\sigma = \frac{1}{j} \sum_{j} \left[I_{tot}^{model}(j) - I_{tot}^{expt}(j) \right]^2 \tag{3.8}$$

where squaring removes any possible errors due to differences in sign. Multiplying through by $\frac{1}{j}$ gives an average per LSF pixel.

There are caveats with LSF: intensities must be scaled to match for both images, otherwise one image will have a heavier weighting on σ than the other. Also, the cluster must be positioned in similar areas of the compared images: we can use the centre of mass (COM) to centre the experimental images, whilst for modelling we use the model cluster COM and centre this with respect to our intensity map before comparison.

3.2.3 *Implementation*

In both the covariance and LSF measures, σ is dependent on the orientation of the 3D theoretical structure used for modelling with respect to the xy-imaging plane. We define, and implement sequentially, the possible cluster rotations using the Tait-Bryan angles θ, ϕ and ψ around the principle axes x-, y- and z- [18], respectively, giving a unique residual $\sigma_{\theta,\phi,\psi}$. We use a floating coordinate system (i.e. where the axes are not fixed). A maximum range of $0 \leq \theta < 2\pi$, $-\frac{\pi}{2} \leq \phi \leq \frac{\pi}{2}$ and $0 \leq \psi < 2\pi$ is needed to cover all the orientations, though this can be decreased for symmetric clusters. We note that should $\phi = -\frac{\pi}{2}$ or $\phi = \frac{\pi}{2}$, the x- and z-axes will be aligned in what is known as "Gimbal Lock" [19], and duplicate orientations of the cluster will be achieved for any values of θ and ψ, so long as $\theta + \psi = c$, where c is a constant.

3.3 Results and Discussion I: Monolayer-Protected Au_{38} Linear Search

Computational calculations of the structure of MP-Au_{38} using density functional theory (DFT) have been performed by Häkkinen et al., Pei et al. and Lopez-Avecedo et al., who identify two energetically low-lying motifs for Au_{38} in the presence of 24 thiolate (SCH_3) ligands (Fig. 3.3): (a) a spherical arrangement, with an Au_{14} core [20], with a maximum aspect ratio of 1.2, and (b) a prolate structure, with a bi-icosahedral core of Au_{23} [21, 22] and aspect ratio ~1.6. For the latter, (b), the arrangement of the thiolate ligands proved critical in matching experimental XRD results with theoretical calculations [22]. Henceforth, the coordinate systems are referred to as MP-Au_{38}(a) and MP-Au_{38}(b), respectively. Both structures show the "divide and protect" morphology associated with monolayer-protected clusters [23], where the cluster is partitioned into a pure core, and an outer layer which interacts

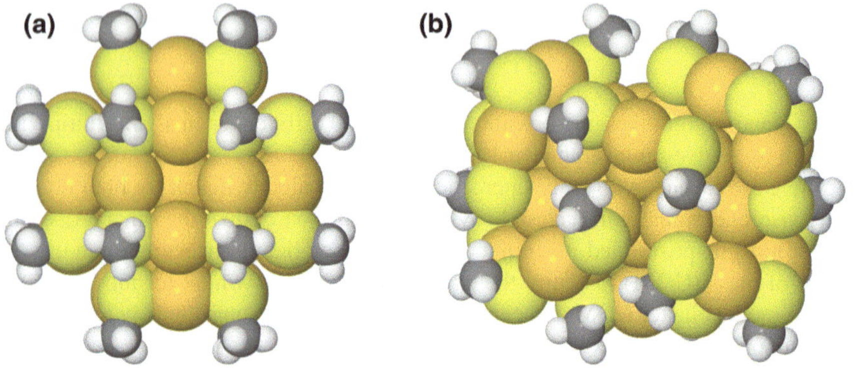

Fig. 3.3 MP-Au_{38} coordinates from DFT-calculations: **a** spherical motif, with an Au_{14} core [20] and **b** a prolate geometry, with a bi-icosahedral core of Au_{23} [21]. Au, S, C and H atoms are represented in *gold*, *yellow*, *grey* and *white*, respectively

strongly with the thiolate ligands. Wang et al. compared their experimental HAADF-STEM images of MP-Au_{38} with the computational structures, statistically associating the prolate geometries observed for the experimental clusters with the bi-icosahedral core cluster in Fig. 3.3b [14].

To test our model, and assess computation time, we looked to perform a more thorough investigation of the orientation of the DFT Au_{38} structures which best match

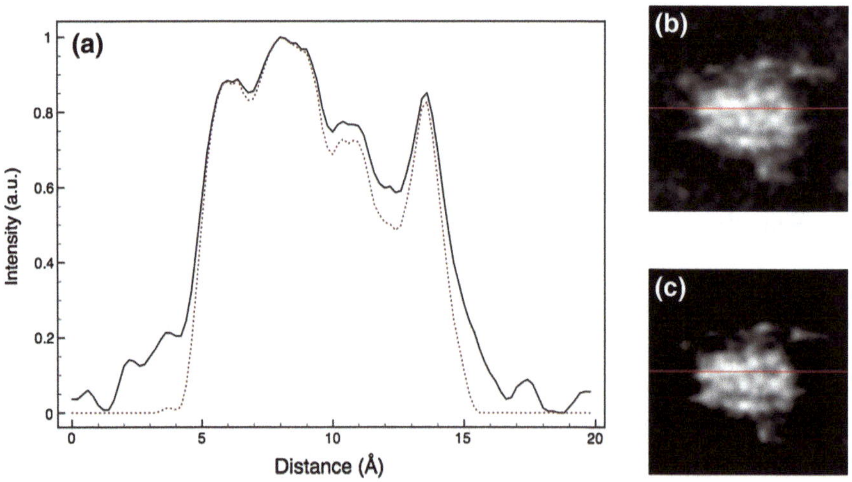

Fig. 3.4 **a** Normalised linear intensity profile from experimental MP-Au_{38} images, provided by Dr. Z. Wang [24], with **b** the original image (*black, solid line*) and **c** the same image with background noise removed (*red, dotted line*). Intensity profiles have been taken horizontally along the *red line* indicated i.e. through the middle of the image

an experimental image, using a linear search through $0 \leq \theta < 2\pi$, $-\frac{\pi}{2} \leq \phi \leq \frac{\pi}{2}$ and $0 \leq \psi < 2\pi$, in 5° steps (197173 calculations). A 2×2 nm experimental image (100×100 pixels) was provided by Dr. Z. Wang (Fig. 3.4) [24].

We simulated on a $\Delta x = \Delta y = 0.2$ Å grid, with the experimentally measurable β value set to 1.7. r_i and Z_i were set to 1.35 Å and 79 for Au, respectively; 1.0 Å and 16 for S, respectively; 0.7 Å and 6 for C, respectively; and 0.25 Å and 1 for H, respectively [25]. Background noise from the amorphous graphite substrate used to support nanoclusters in the experimental work interferes with images and so calculations were performed both with the original image and with background noise removed using a threshold level of intensity 50 (Fig. 3.4c).

3.3.1 $\alpha = 1$

STEM image simulations were calculated with $\alpha = 1$; results for FEs using LSF and covariance are shown in Table 3.1. Images of the optimal orientations are given in Figs. 3.5 and 3.6, along with linear intensity profiles. The linear profiles show the difficulty in identifying atomic columns in these structures when compared to other high-symmetry arrangements [16].

For MP-Au_{38}(a), use of the LSF FE gives an optimal orientation of $(\theta, \phi, \psi) = (180°, -65°, 185°)$ (Fig. 3.5b), with one other similar orientation $(\theta + 5°, \phi + 5°)$ in the lowest 10 $\sigma_{\theta,\phi,\psi}$ values; the same is true for the covariance FE where the optimal orientation is $(180°, -45°, 275°)$, but several rotated orientations (θ, ϕ, or $\psi \pm 90°$) are competitive. The mirrored/rotated orientations are competitive as the MP-Au_{38}(a) structure has O_h symmetry in its Au_{38} core. In the presence of the thiolate ligands this becomes psuedo-O_h, as the ligands do not align symmetrically. This leads to a frustrated function landscape, with many competitive, but disconnected, minima.

Table 3.1 Results, simulating MP-Au_{38} structures using $\alpha = 1$, for the optimal orientation which minimises the chosen FE method, taken using an exhaustive search with 5° steps over the orientational space

Structure	Background threshold	FE method	Minima (θ, ϕ, ψ)	$\sigma_{\theta,\phi,\psi}$
MP-Au_{38}(a)	0	LSF	180°, 65°, 185°	782.841
	50	LSF	270°, −45°, 185°	870.181
	0	Covariance	180°, −45°, 275°	2340.23
	50	Covariance	180°, −45°, 270°	1994.94
MP-Au_{38}(b)	0	LSF	120°, 5°, 265°	677.42
	50	LSF	10°, −25°, 320°	539.687
	0	Covariance	300°, 45°, 160°	2214.74
	50	Covariance	300°, 45°, 160°	1998.68

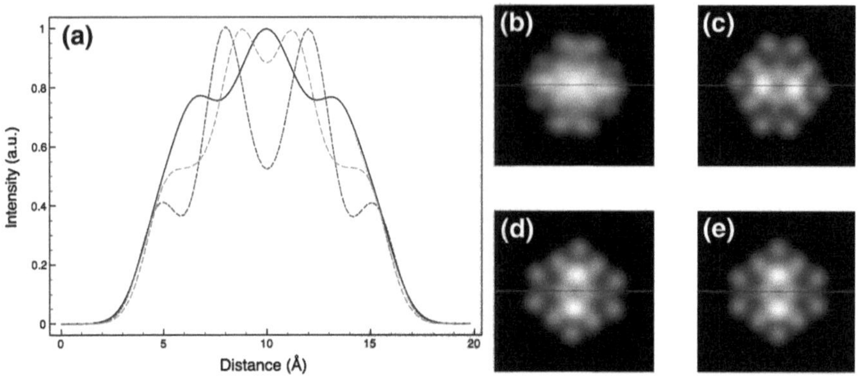

Fig. 3.5 **a** Normalised linear intensity profiles for optimal orientations of MP-Au_{38}(a), as given in Table 3.1. Profiles given where (θ, ϕ, ψ) is: **b** $(180°, -65°, 185°)$ (*black, solid line*); **c** $(270°, -45°, 185°)$ (*red, dotted line*); **d** $(180°, -45°, 275°)$ (*blue, dashed line*); and **e** $(180°, -45°, 270°)$ (*green, long-dashed line*). Intensity profiles have been taken horizontally along the *red line* indicated i.e. through the middle of the image

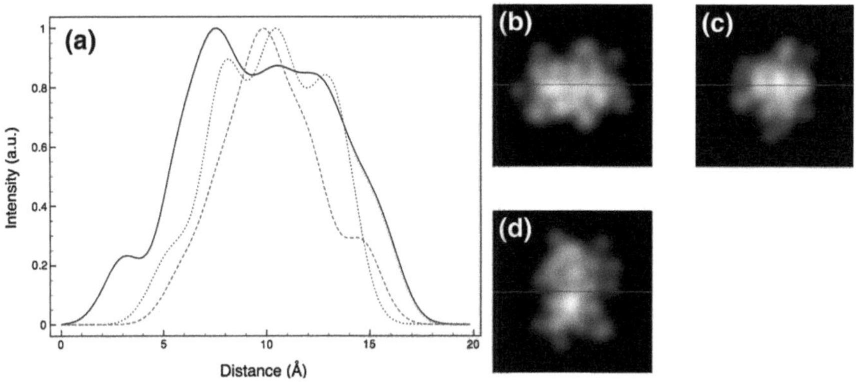

Fig. 3.6 **a** Normalised linear intensity profiles for the optimal orientations of MP-Au_{38}(b), as given in Table 3.1. Profiles given where (θ, ϕ, ψ) is: **b** $(120°, 5°, 265°)$ (*black, solid line*); **c** $(10°, -25°, 320°)$ (*red, dotted line*); and **d** $(300°, 45°, 160°)$ (*blue, dashed line*). Intensity profiles have been taken horizontally along the *red line* indicated i.e. through the middle of the image

In the case of MP-Au_{38}(b), using the LSF FE gives an optimal orientation of $(\theta, \phi, \psi) = (120°, 5°, 265°)$ (Fig. 3.6b), with all of the lowest 10 $\sigma_{\theta,\phi,\psi}$ values being close in orientation ($\theta \pm 10°$, $\phi \pm 10°$, $\psi \pm 10°$). For the covariance FE the orientation $(300°, 45°, 160°)$ is found to be optimal, with the lowest ten minima again being closely related ($\theta \pm 5°$, $\phi \pm 5°$, $\psi \pm 5°$).

Removal of background noise results in slight changes to the optimal (θ, ϕ, ψ) orientation found for MP-Au_{38}(a) using the LSF FE. No change in orientation is seen for MP-Au_{38}(b) using the LSF FE, or for the covariance FE where the $\sigma_{\theta,\phi,\psi}$ minima prove consistent for both structures.

Removal of Ligands (Au_{38}-Core)

The contribution to our modelled STEM image is dominated by heavier elements due to the dependence on Z_i: in our current calculations a gold atom has fifteen times the intensity of a sulphur atom. The calculation time for our STEM models scales linearly with the number of atoms (N) in the theoretical cluster; therefore considerable speed up is achieved following removal of the ligands from the structure coordinates. With this in mind we re-ran our linear searches with just the Au atoms; the results are shown in Table 3.2 and displayed graphically in Figs. 3.7 and 3.8.

As with the presence of the thiolate ligands for MP-Au_{38}(a), many rotated and mirrored orientations are competitive due to the O_h symmetry of the Au_{38}-core(a) structure, which was now treated explicitly (Fig. 3.7). Therefore, no common values of θ, ϕ and ψ ($\pm 10°$) are noted at a frequency greater than twice for the 10 lowest

Table 3.2 Results, simulating Au_{38}-core structures using $\alpha = 1$, for the optimal orientation which minimises the chosen FE method, taken using an exhaustive search with 5° steps over the orientational space

Structure	Background threshold	FE method	Minima (θ, ϕ, ψ)	$\sigma_{\theta,\phi,\psi}$
MP-Au_{38}(a)	0	LSF	185°, −65°, 190°	788.965
	50	LSF	270°, −50°, 185°	848.019
	0	Covariance	180°, −45°, 275°	2309.48
	50	Covariance	180°, −45°, 270°	2089.46
MP-Au_{38}(b)	0	LSF	125°, 5°, 265°	699.82
	50	LSF	10°, −25°, 320°	522.894
	0	Covariance	300°, 45°, 160°	2206.29
	50	Covariance	300°, 45°, 160°	1994.94

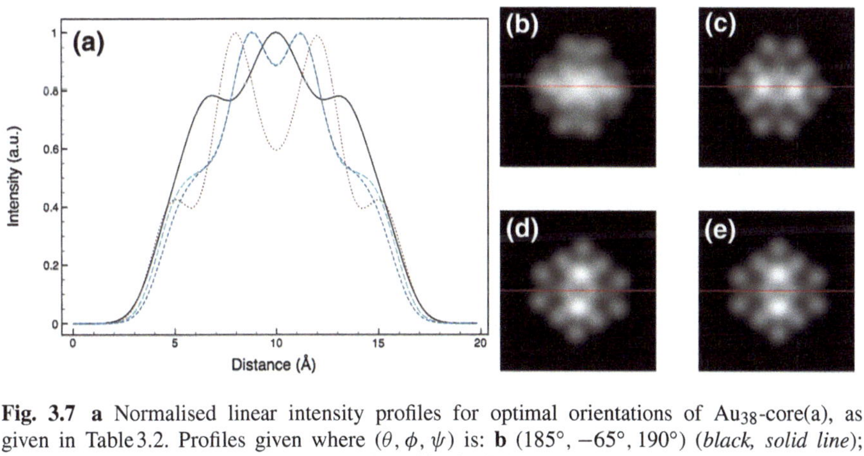

Fig. 3.7 **a** Normalised linear intensity profiles for optimal orientations of Au_{38}-core(a), as given in Table 3.2. Profiles given where (θ, ϕ, ψ) is: **b** (185°, −65°, 190°) (*black, solid line*); **c** (270°, −50°, 185°) (*red, dotted line*); **d** (180°, −45°, 275°) (*blue, dashed line*); and **e** (180°, −45°, 270°) (*green, long-dashed line*). Intensity profiles have been taken horizontally along the *red line* indicated i.e. through the middle of the image

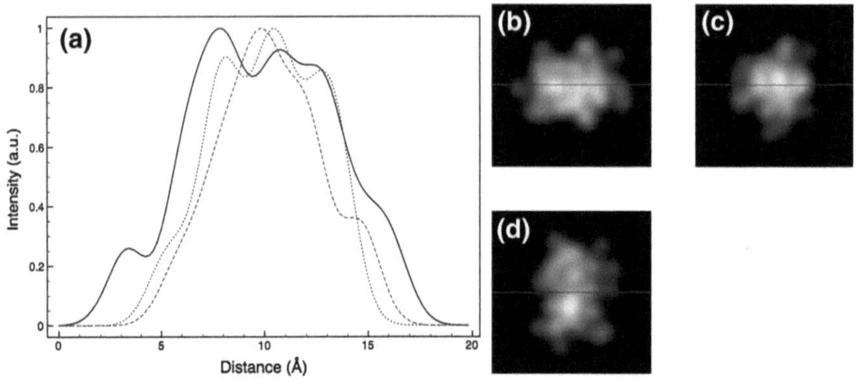

Fig. 3.8 **a** Normalised linear intensity profiles for optimal orientations of Au_{38}-core(b), as given in Table 3.2. Profiles given where (θ, ϕ, ψ) is: **b** $(125°, 5°, 265°)$ (*black, solid line*); **c** $(10°, -25°, 320°)$ (*red, dotted line*); and **d** $(300°, 45°, 160°)$ (*blue, dashed line*). Intensity profiles have been taken horizontally along the *red line* indicated i.e. through the middle of the image

$\sigma_{\theta,\phi,\psi}$ values gathered from model comparison to the original image, though some mirror orientations are found to be close to the minimum (i.e. $\theta + 180°$ or $\psi + 180°$). Introduction of a background threshold to remove background interference does not improve the frustrated landscape for the covariance FE, whilst the LSF FE shows a more consistent result with 5 of the lowest 10 minima being related (θ or ϕ or $\psi \pm 10°$) to the global minimum (GM).

For Au_{38}-core(b), use of the LSF FE gave 5 out of the lowest 10 results with the orientation $\theta = 125° \pm 5°$, $\phi = 5° \pm 5°$ and $\psi = 265° \pm 5°$ (Fig. 3.8a), whilst adding a background threshold resulted in all of the lowest 10 results being the same within θ or ϕ or $\psi \pm 10°$. This high consistency is similar to our results with the inclusion of the SCH_3 thiolate ligands. The covariance FE also gives consistent results, with 6 of the lowest 10 (θ, ϕ, ψ) orientations being within $\pm 10°$ of each other; use of a background threshold increases this to 7 out of the top 10.

Overall, for both Au_{38}-core(a) and (b) our minimum orientations are consistent with calculations performed with the thiolate ligands present, with one exception for MP-Au_{38}(a) which differs by 5° in θ and ϕ. As with calculations performed including the thiolate ligands, LSF FEs prove more sensitive to the removal of background noise compared to using the covariance FE.

A normalised linear profile from the MP-Au_{38}(b), with optimum orientation from the LSF FE of $(\theta, \phi, \psi) = (120°, 5°, 265°)$ is given in Fig. 3.9, compared to experimental images. The linear profiles do not match the experimental image particularly well, despite qualitatively appearing similar to the experimental images. Also in Fig. 3.9 we plot the normalised linear profiles with and without inclusion of the SCH_3 ligands for the optimal MP-Au_{38}(b) and Au_{38}-core(b) orientations, noticing that subtle differences are made to peak locations and intensity despite their relative similarity to the naked eye (Figs. 3.6b and 3.8b).

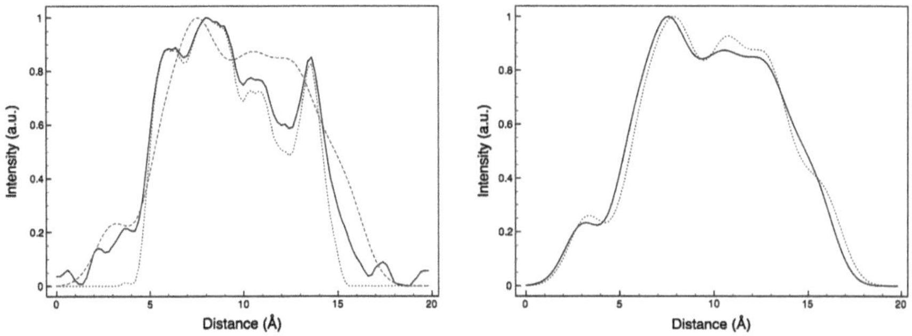

Fig. 3.9 Comparison of normalised linear intensity profiles. *Left* The original HAADF-STEM image (*black, solid line*, taken from 3.4b); original HAADF-STEM with background noise removed (*red, dotted line*, taken from Fig. 3.4c) and modelled STEM image of MP-Au_{38}(b), oriented at $(\theta, \phi, \psi) = (120°, 5°, 265°)$ (*blue, dashed line*, taken from Fig. 3.6b). *Right*: Optimal orientations of MP-Au_{38}(b) (*black, solid line*) compared to Au_{38}-core(b) (*red, dotted line*, taken from Fig. 3.8b). Intensity profiles have been taken horizontally along the *red line* indicated on original images i.e. through the middle of the images

Quantitatively, for all calculations performed with $\alpha = 1$ the minimum $\sigma_{\theta,\phi,\psi}$ values are lower for MP-Au_{38}(b) than MP-Au_{38}(a), using both the LSF and covariance FE. For MP-Au_{38}(a) use of background noise subtraction with the LSF FE results in an increased minimum value for $\sigma_{\theta,\phi,\psi}$, a result not duplicated for other conditions and perhaps indicative that the MP-Au_{38}(a) structure does not match the experimental image well.

3.3.2 $\alpha = 1.5$

To gauge the sensitivity of results to variation in the gaussian exponent (α) used in our STEM model, and the influence this has on results over a landscape search, the linear search was repeated using $\alpha = 1.5$. A larger α value will sharpen the atoms in our simulated images, potentially leading to more complex function landscapes. Initial results are given in Table 3.3, with images and linear profiles provided in Figs. 3.10 and 3.11.

MP-Au_{38}(a) proves to have a frustrated landscape, due to the underlying symmetry, with no similar ($\pm 5°$) values of θ, ϕ and ψ occurring more than twice in the 10 most optimal orientations, with a solitary exception noted using the covariance FE coupled with an image with the background noise removed, where 3 values where found within $\psi \pm 5°$ of each other. In contrast, for MP-Au_{38}(b) using the LSF FE results in all the optimal (θ, ϕ, ψ) values were within $\pm 10°$ of each other. Using the covariance FE on MP-Au_{38}(b), 5 of the 10 optimal (θ, ϕ, ψ) orientations prove to be similar before background noise is subtracted, and this is further improved by the removal of background noise from the experimental image.

Table 3.3 Results, simulating MP-Au_{38} structures using $\alpha = 1.5$, for the optimal orientation which minimises the chosen FE method, taken using an exhaustive search with 5° steps over the orientational space

Structure	Background threshold	FE method	Minima (θ, ϕ, ψ)	$\sigma_{\theta,\phi,\psi}$
MP-Au_{38}(a)	0	LSF	185°, −65°, 190°	953.913
	50	LSF	180°, −65°, 185°	908.715
	0	Covariance	180°, −45°, 275°	1788.42
	50	Covariance	180°, −45°, 270°	1617.89
MP-Au_{38}(b)	0	LSF	120°, −5°, 265°	769.929
	50	LSF	10°, −20°, 315°	541.456
	0	Covariance	340°, 45°, 285°	1811.86
	50	Covariance	300°, 45°, 160°	1631.71

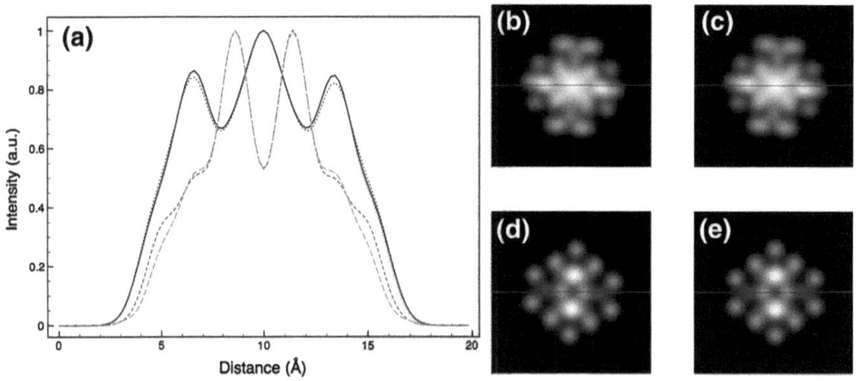

Fig. 3.10 **a** Normalised linear intensity profiles for optimal orientations of MP-Au_{38}(a), as given in Table 3.3. Profiles given where (θ, ϕ, ψ) is: **b** (185°, −65°, 190°) (*black, solid line*); **c** (180°, −65°, 185°) (*red, dotted line*); **d** (180°, −45°, 275°) (*blue, dashed line*); and **e** (180°, −45°, 270°) (*green, long-dashed line*). Intensity profiles have been taken horizontally along the *red line* indicated i.e. through the middle of the image

Our results prove remarkably similar to those for $\alpha = 1$ in Sect. 3.3.1, with most optimal (θ, ϕ, ψ) orientations being within 5° of those results. The only exception is seen for MP-Au_{38}(a) where, using the LSF FE and background noise subtraction, a different results is seen: $(\theta, \phi, \psi) = (180°, 65°, 185°)$ compared to (270°, −50°, 185°) previously for $\alpha = 1$. However, we know the landscape of MP-Au_{38}(a) is complicated due to the symmetry of the structure, and the 2nd optimal orientation in this case is similar $(\phi + 15°)$ to the orientation seen for $\alpha = 1$.

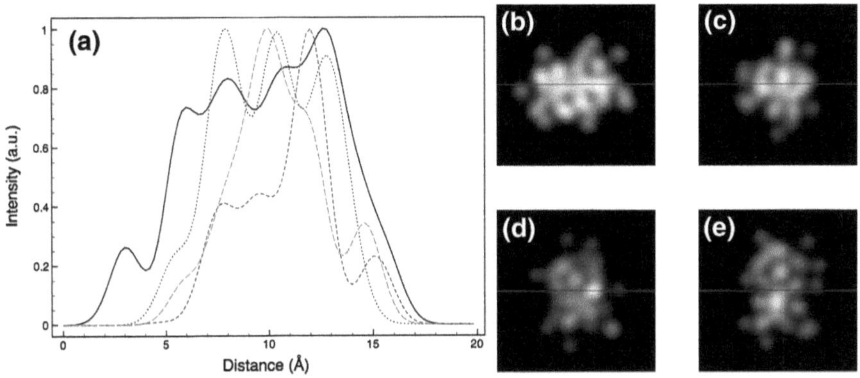

Fig. 3.11 **a** Normalised linear intensity profiles for optimal orientations of MP-Au_{38}(b), as given in Table 3.3. Profiles given where (θ, ϕ, ψ) is: **b** $(120°, -5°, 265°)$ (*black, solid line*); **c** $(10°, -20°, 315°)$ (*red, dotted line*); **d** $(340°, 45°, 285°)$ (*blue, dashed line*); and **e** $(300°, 45°, 160°)$ (*green, long-dashed line*). Intensity profiles have been taken horizontally along the *red line* indicated i.e. through the middle of the image

Removing Ligands from DFT-Structures

We removed the thiolate (SCH_3) coordinates from our theoretical models to see if we are able to produce similar results as calculated with the ligands included, based on the presumption that Au contributes strongly to the overall intensity. Results for a linear search comparing theoretical Au_{38}-core structures with an experimental MP-Au_{38} STEM image are given in Appendix B.1, along with comparative images and linear intensity profiles.

Quantitative analysis shows us that the only optimal orientation which changes, compared to the inclusion of the SCH_3 thiolate ligands is for MP-Au_{38}(a), using the LSF FE and zero background subtraction. Furthermore, we notice that for all calculations where $\alpha = 1.5$, the minimum $\sigma_{\theta,\phi,\psi}$ value decreases for both the LSF and covariance FE when a background threshold is used to remove noise from the image. Both with and without ligands attached, the MP-Au_{38}(b) structure gives a $\sigma_{\theta,\phi,\psi}$ value lower than MP-Au_{38}(a) with the use of the LSF FE, with the opposite being true—i.e. MP-Au_{38}(a) returns lower values than MP-Au_{38}(b)—with the use of the covariance FE.

3.3.3 Discussion

We have systematically searched orientational space (θ, ϕ, ψ) using kinematically modelled STEM images statistically compared to an experimentally measured image. Our observations show good agreement to the previous results, where the elongated structure of MP-Au_{38}(b) was likened to the prolate nanocluster projections seen through the HAADF-STEM [14]. Qualitatively it is easier to match the LSF FE results with the experimental image than the results produced using the covariance FE, and this is something which needs further consideration. Also, the lack of correlating results once the background noise is removed may be associated with a loss of real features with noise subtraction—something which needs careful attention in future work.

Several method improvements can be identified from this work: broadly they separate into improvement of the HAADF-STEM model, and improvement of the search through orientational space.

The experimental image for MP-Au_{38} we have compared to modelled solutions is an image snapshot, and this must be remembered when reviewing results. The small clusters are both atomically mobile, with rearrangements occurring continuously under the electron beam, and molecularly mobile on the graphite substrate, with agglomerations regularly forming [14]. This mobility means that identifying high symmetry motifs, as has been done previously [1], is challenging. Possible solutions to this problem include the addition of a stochastic method to rearrange clusters such that their atomic arrangement matches the experimental images better, though this is artificial in nature. An alternative would be to use thermal molecular dynamics simulations to give a database of rearranged structures with which to compare to the experimental image, with the additional inclusion of atomic vibrational effects within our model.

Whilst the linear search method is thorough, systematic searches are computationally inefficient and this is something which requires direct addressing. More suitable is the implementation of a minimisation method, both local and global, in order to accelerate the searching of the 3D hypersurface. Also, with the advent of modern computers it is possible to harness multiple central processing units (CPUs) simultaneously to share out calculative task, and this too can significantly improve real-time calculation speed. Improvements to the search method are the focus of the following section, and both of the previous two points—minimisation and parallelisation—will be addressed.

3.4 Methodological Improvements

From Sect. 3.2.3 we have shown the potential use of a simple kinematic STEM model, though the search is still unnecessarily computationally laborious. Naively performing an exhaustive linear search with 5° steps is computationally expensive ($72 \times 37 \times 72 = 191{,}808$ FEs) and does not cover the search area in great details; a search with 1° steps would require $23,457,600$ FEs. Introduction of a global minimisation routine could be used to improve search efficiency.

Global optimisation is a search technique for optimising a numerical function with many degrees of freedom. In our case, we have three variables θ, ϕ and ψ; and by changing them from random starting values we can navigate the multidimensional hypersurface, travelling towards negative gradients in order to optimise $\sigma_{\theta,\phi,\psi}$.

Different search methods favour different kinds of problems: stochastic global minimisation methods which explore localised areas, such as Metropolis Monte Carlo, are suited to funnelled landscapes. For more frustrated landscapes wider search methods are more useful, such as genetic algorithms (GAs) [26], and in some cases coupling of heuristic and stochastic search methods has proven successful [27].

3.4.1 Genetic Algorithms

GAs are search methods for function minimisation based on natural evolution, as first documented by Charles Darwin [28]. Operators analogous to those observed in nature are used: mating, mutation and natural selection, to explore multidimensional hypersurfaces in search of a function solution. GAs belong to the class of evolutionary algorithms, which also include evolution strategies, differential evolution and genetic programming [29].

The application and use of a GA is problem specific and as such the operators involved (mating, selection, mutation, etc.) are optimised to the problem, depending on the level of correlation within the search results. The wide range of numerical problems in chemistry allows varied application of the search technique, most commonly when matching experiment with theory [30]. Example applications include: the determination of molecule structure from spectroscopic methods, including NMR (biomolecules) [31]; XRD (crystal structures, both powder and single crystal) [32]; and UV-Vis [33] spectra; the design and docking of drug molecules [34]; and the optimisation of chemical processes [35]. Predictions are also made of molecule structure, from nanoclusters [13, 36–38] through to protein folding [39, 40].

Terminology

In principle, a GA can be applied to any multidimensional function where the variables to be optimised ("genes"), with distinct values ("alleles"), can be regarded as a string ("chromosome"). Each string is a a trial solution to a problem, and a **population** can be regarded as a collection of these trial solutions. It is normal to take a starting population and evolve it over numerous **generations** until a **convergence criterion** is met, or the maximum number of generations is reached.

Members of the **initial population** are normally generated at random, however it can be beneficial to bias the initial population using prior knowledge of the problem being addressed [13].

Each population member is then assessed for its **fitness** (f_i). This important concept ranks the members of a population using their proximity to the desired problem solution as the ranking variable (i.e. lowest for minimisation, or highest for maximisation), and is very important in determining whether a population member is accepted for mating and/or survival from one population generation to the next. If the upper and lower bounds of the search area are known then **absolute fitness** can be used; however this is uncommon in general minimisation searches. Ranking the population members relative to each other, known as **dynamic fitness evaluation**, is used for GAs where the final solution is unknown.

The ranking is achieved by normalising the fitness value such that:

$$\rho_i = (V_i - V_{min})/(V_{max} - V_{min}) \tag{3.9}$$

where V_{min}, V_{max} and V_i represent the current function minimum, maximum and ith value, respectively. Fitness functions implementable include the **linear** fitness function:

$$f_i = 1 - 0.7\rho_i \tag{3.10}$$

the **exponential** fitness function:

$$f_i = e^{-\alpha\rho_i} \tag{3.11}$$

where α is usually set to 3; and the **hyperbolic tangent** fitness:

$$f_i = \frac{1}{2}[1 - \tanh(2\rho_i - 1)]. \tag{3.12}$$

Once all members have had their fitness evaluated, **selection** is required to determine the members to be put forward for mating. Numerous selection processes exist, the two most common being "roulette wheel" and "tournament" selection. For **roulette wheel** selection a string is selected at random, and if its fitness (f_i) is greater than a randomly selected number between 0 and 1 (i.e. $f_i > R[0, 1]$) it is put forward for mating, otherwise another string is selected and the process repeated, until a mating pair is created. The analogy with a roulette wheel can be imagined if one thinks of a roulette wheel with varying size slots for each population member, dependent on f_i: strings with a higher f_i having a larger width slot, and thus greater chance of selection. As an alternative, **tournament** selection chooses a pool of strings at random, of which the two members with the highest fitness are put forward for mating.

The selected strings are used for **mating**, or **crossover**, allowing offspring to be formed with genetic properties of both parents (i.e. inheritance). Mating operators are implemented at a predefined frequency, normally dependent on the population size. Several possibilities are available for crossover: **one-point crossover** involves randomly selecting a cutting-point of the parent strings, cutting the strings at this same point and then splicing them together to form the offspring with the complimentary properties of its parents (Fig. 3.12a). **Two-point crossover** is an extension of this principle with two cutting-points randomly identified; thus an offspring is formed with a segment of one parent's genes sandwiched between the genes of the other parent. This concept can be extended to three-point crossover and more if desired, though this is of greater functional use for large strings. At the extreme end of this methodology is **uniform crossover**, where each gene for the offspring is selected

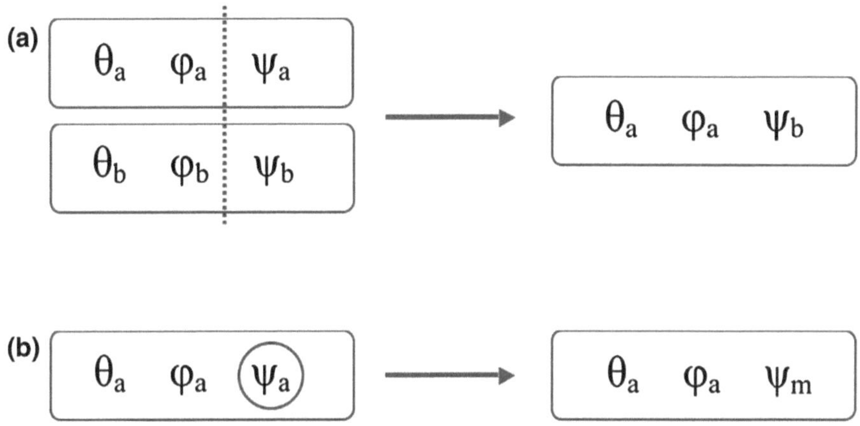

Fig. 3.12 **a** One-point crossover, with the cutting point for our "chromosome" marked with a *red, dashed line*. The values of θ_a and ϕ_a have been combined with ψ_b. The offspring is given on the *right hand side*. **b** Mutation, with one allele (ψ) highlighted on the *left* for mutation, and altered to a random value ψ_m

randomly from one of the two parents. (i.e. potentially an N-point crossover for a string of length N).

Whilst crossover leads to the exchange of genetic information, no new material is introduced, which can lead to stagnation of the population, i.e. convergence on a non-optimal solution. **Mutation** deals with this problem by introducing new random property values to the strings, increasing diversity (Fig. 3.12b). For **static mutation** the new variable is chosen completely at random, within the limits of the search area, whilst for **dynamic mutation** a small, random value change is made to a variable: In the latter case consideration is made to the number of completed generations without change of the best solution, with the mutation vector decreasing as this number increases. Mutation can be implemented on the parents (PM) or offspring (OM) in the population in each generation. Mutation operators are usually implemented at a predefined frequency, similar to the mating operator.

Finally, the process of **natural selection** completes the evolution of one population set. This selection is conducted by using the fitness of each population member, and many possible variations of selection are possible. The most commonly used is **elitist** selection, where only the best population members from the original population, offspring and mutants pass into the new population.

Once the cycle is complete the **convergence criterion** is referred to and, if not met, then the search returns to the "fitness" stage, whilst if the criterion is met the algorithm will terminate. The convergence criterion could be the number of generations without improvement of the best solution, reaching a pre-defined solution threshold, or just reaching the maximum number of generations permitted.

3.4.2 Minimisation

Local Minimisation

Due to the dependence of $\sigma_{\theta,\phi,\psi}$ on cluster orientation, analytical derivatives are not available for local minimisation, and so numerical methods are used, as outlined below, in a "steepest-descent" method. Numerical minimisation of the function is possible by optimising the orientation variables θ, ϕ and ψ independently or together; these are known as univariate and multivariate search methods, respectively.

A **univariate search** constitutes the linear variation of only one variable at a time within the search vector. After one variable is optimised, the search proceeds to the next variable, and so forth, looping back to the beginning when it reaches the end. Optimisation of the variables is complete when the function has reached the lowest available point on the hypersurface: the function has been locally minimised.

Powell's conjugate gradient descent method, more commonly known as **Powell's method**, introduces an additional search vector to the univariate search which is a combination of the successful search moves in the univariate phase [41]. The new search vector is added to the end of the univariate search vector list, with the advantage over a univariate search being that it involves diagonal movements as well as linear on the hypersurface. Again, optimisation is complete once all variables have values points such that the function is at the lowest locally available point on its hypersurface.

Multivariate searches involves variation of all variables within the search vector at the same time, and is only useful when the system in question consists of only a few variables as the computational costs scale many magnitudes larger than univariate searches: In our case for each multivariate step we require 26 FEs, compared to 2 FEs for univariate searching. The obvious advantage of the multivariate search is that it allows a more varied search path to minimisation.

3.4.3 Combining Global and Local Minimisation

We have shown that naively performing an exhaustive linear search is computationally expensive, therefore we introduce minimisation to improve search efficiency. By changing θ, ϕ and ψ from starting values we can navigate the multidimensional hypersurface, travelling towards negative gradients in order to locally minimise $\sigma_{\theta,\phi,\psi}$. However, performing one local minimisation of the cluster orientation does not guarantee locating the global minimum.

Therefore we implement a global minimisation method in our work: more specifically a GA. It has been shown that coupling global and local minimisation can improve search routines [13], in what is known as a Lamarckian search method. GA calculations can be performed in two regimes: a "Darwinian" GA (D-GA) has a non-adapting population from one generation to the next (i.e. no local minimisation),

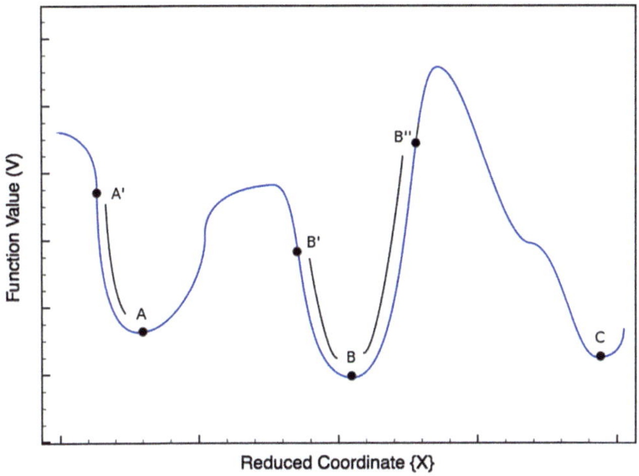

Fig. 3.13 Example of a function landscape (*blue line*). The global minimum is point B, whilst other local minima lie at points A and C. Points A', B' and B" represent non-minimum points on the hypersurface; in a Darwinian-GA these would points would be carried forward to the fitness function stage as they are, whilst in the Lamarckian-GA these points would be minimised (as illustrated by *black lines*), with point A' minimising to point A, and points B' and B" minimising to point B

whilst a "Lamarckian" GA (L-GA) which locally-minimises the function for each population member, at each step [42], reducing the landscape to step-like features (Fig. 3.13). In the following work both are tested, and will be referred to henceforth with the FE-type appended to their abbreviation: the four combinations being D-LSF, D-covariance, L-LSF and L-covariance.

3.5 Development and Implementation

3.5.1 Development

Storing Search Paths

The search route can be saved, so that should any route be repeated in our minimisation we do not have to recalculate $\sigma_{\theta,\phi,\psi}$ but instead retrieve it from memory (historical evaluation, HE). HEs are stored in vector arrays, which are cleared between each GA run. Duplicate points, such as in the case of "Gimbal Lock", have the same value for $\sigma_{\theta,\phi,\psi}$ stored in all duplicated positions preventing unnecessary FEs (akin to a Tabu Search [43]).

Choosing Random Values for θ, ϕ and ψ

All random numbers are generated using the *ran3* subroutine documented elsewhere [41]: varying the initial random number seed ensures no search path is duplicated during testing. For θ and ψ, random values can be picked to give a uniform distribution of initial points; however for ϕ this would give a clustering of starting points at the spherical poles ($\frac{\pi}{2}$ and $-\frac{\pi}{2}$). Therefore the random points are chosen as:

$$\theta = 2\pi u \tag{3.13}$$

$$\phi = \cos^{-1}(2v - 1) \tag{3.14}$$

$$\psi = 2\pi w \tag{3.15}$$

where the different random numbers u, v and w are defined as $R[0, 1]$. In our searches, cluster symmetry is not used to limit θ, ϕ and ψ, i.e. $0 \leq \theta < 2\pi$, $-\frac{\pi}{2} \leq \phi \leq \frac{\pi}{2}$ and $0 \leq \psi < 2\pi$, however periodic boundary conditions are used on complete rotations (i.e. θ: $2\pi = 0$). We use 1° steps to separate the rotations throughout.

Parallelisation

Due to the nature of the search algorithm, several FEs are required at each step. FEs are independent of each other and parallelisation of the code across multiple processors on one node using Open Message Passing (OMP) can be applied trivially; it is necessary to use shared-memory parallelisation so that checking the HEs is available to multiple processors (Fig. 3.14). We can also parallelise across nodes using Message Passing Interface (MPI) for multiple structures/images in a

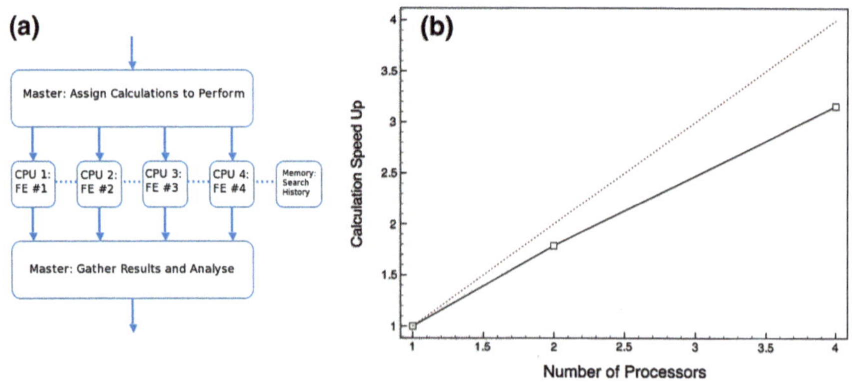

Fig. 3.14 OMP Implementation: **a** Schematic flow diagram for the parallelisation involving the splitting up of local minimisation calculations over multiple processors. **b** Benchmarking calculations for the implemented OMP method. Actual results (*black, solid line*) are compared to idealised results (*red, dotted line*)

"task-farming" manner, leading to a hybrid MPI/OMP system. Implementation of these parallelisation methods is shown to be successful and used henceforth.

3.5.2 *Implementation*

Benchmarking calculations have been performed for the high-symmetry 309-atom Icosahedral (Ih) and Ino-Decahedral (I-Dh) Au clusters (Fig. 3.2), previously compared to experimental images [1, 16]. Structural minimisation was performed using the Gupta many-body atomistic potential [12] with previously defined parameters [44].

Artificial model images were generated as a model solution to the problem, using the predefined Ih or I-Dh cluster coordinates, and then the efficiency of the GA was tested by searching for the correctly matching orientation(s) to this reference data. All model images were calculated on a 30×30 Å grid, with point spacing of 0.5 Å (= 3,600 points). Model parameters of $\alpha = 1.4$ and $r_{Au} = 1.35$ Å [45] were used (Fig. 3.2). For monometallic systems, like the one used here to test the system, the value of β has no influence on results.

Calculations were carried out on the University of Birmingham BlueBEAR supercomputer, which has 384 dual-core worker linux nodes, each with two 2.6 Ghz AMD processors and 8 Gb memory, and a pool of over 150 Tb of storage space [46]. Calculations were repeated 100 times for each search configuration during the parameter testing phase.

3.6 GA Parameterisation and Testing

Parameterisation is necessary with any application of the GA search method due to its problem dependent nature. There are two parameter-types which need optimising for a GA: the types of function, and the frequency of the genetic operator functions (i.e. mating and mutation). In this work we will refer to the frequency as a percentage of the starting population size, with maximum value 100 %.

Initial parameterisation of function types were conducted on an Ih, using L-GA. Local minimisation was performed using a multivariate method. We found that convergence (given as a % of the 100 GA runs which converge to a correct solution) was little affected by the choice of **parent selection** (roulette 88 % vs. tournament 91 %), **mutation type** (static 89 % vs. dynamic 82 %) and **fitness type** (exponential 89 % vs. linear 81 % vs. tanh 86 %).

In the following discussion, unless otherwise stated: roulette selection is used for parent selection, using an exponential fitness function; mating is conducted using uniform crossover; mutation is performed dynamically on population parents (PM); population selection is performed using an elitist method between generations, with a convergence criterion of 8 generations without change of the best solution; population sizes for D-GAs are 1,000 members, and for L-GAs 40 members.

The frequency of mutation and mating defines the best-suited search algorithm-type. We have studied these numerical factors individually and methodically. A typical value of the mating frequency is 80 % of the population (i.e. for a population of 100, 80 offspring will be generated), and mutation frequency is 20 %, and these values are used unless otherwise specified.

3.6.1 Darwinian Genetic Algorithm (D-GA)

Icosahedra

Convergence rates to the GM, for GAs searching for the global optimum values of θ, ϕ and ψ for a 309-atom Ih cluster, are shown in Fig. 3.15, with varying frequencies of the mating and mutation (PM and OM) genetic operators (data tables are available in Appendix B.2). Random search calculations, using random start-points and no global or local minimisation method, yielded no results for the number of FEs required to locate a minima listed (Table 3.4) despite the high-symmetry of the Ih cluster, indicating the sparsity of minima in this problem.

For the D-LSF GA (Fig. 3.15a) a positive correlation is visible between the increasing use of genetic operators and improved convergence. The convergence rate is most affected by PM (2 % → 76 % as the operator frequency increases from 0 % → 100 %, respectively); the improvement in convergence is double that seen for mating and OM. A similar trend is seen for D-covariance GA (Fig. 3.15b), where the increased use of PM has a profound affect on convergence (8 % → 87 %), again approximately double the improvement seen for increasing frequency of mating and OM. In both these D-GAs the frequency of the mating has less influence than mutation on convergence.

Figure 3.16 shows the average FEs and HEs required for calculations with variation of the PM frequency, from 0 to 100 %, that have successfully converged to a GM. We have chosen PM as this initially shows the greatest effect on the convergence rate of the genetic operators tested. The average number of generations before the GM is encountered (FEN) during the search is also given (Fig. 3.16c). Data tables, outlining minimum, average and maximum number of FEs, HEs and FENs for converged

Table 3.4 Benchmark tests using random, non-global, searches. Calculations were performed without minimisation enabled, over a time period of 100 h each

Calculation type	FE type	FEs	Minimum found	Ratio
Point	LSF	1,724,042	0	–
	Covariance	1,723,750	0	–

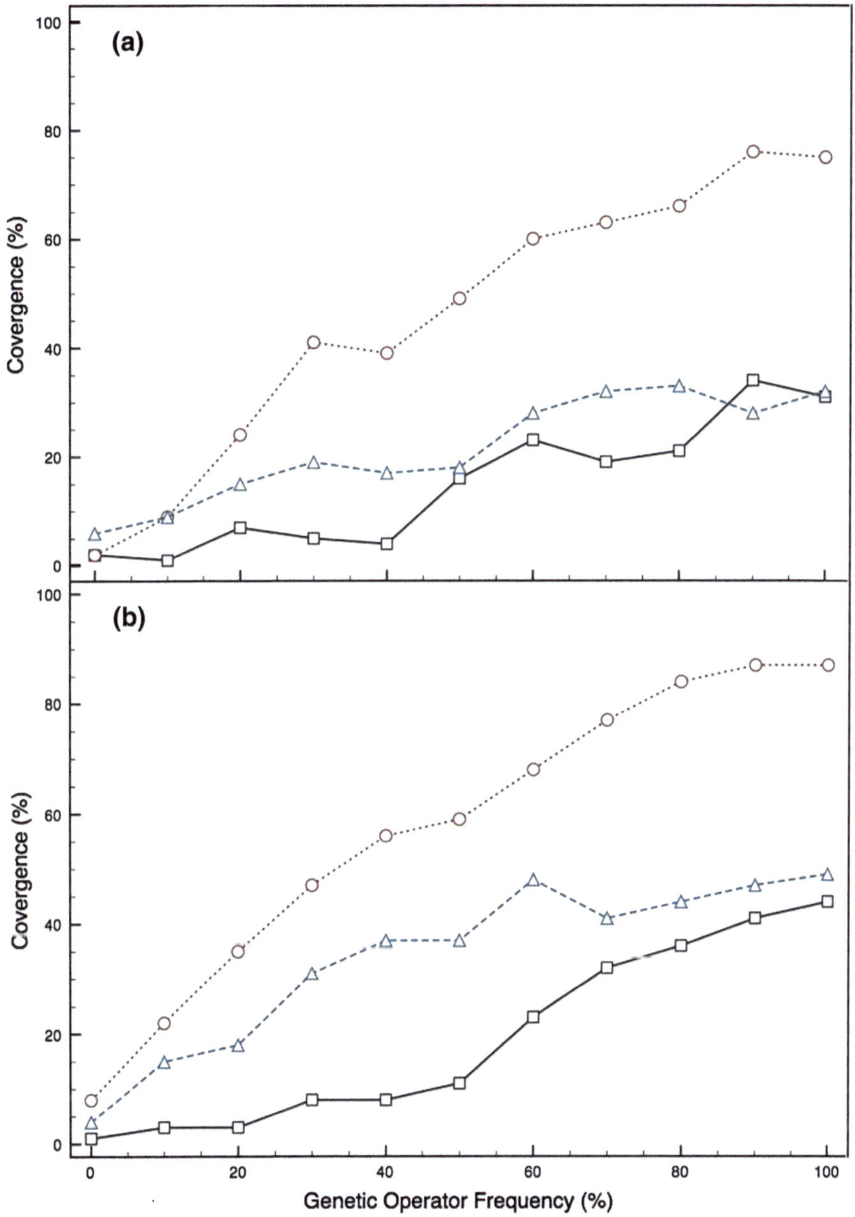

Fig. 3.15 For Ih structure: **a** Convergence rate for the minimised orientation of Ih to model solution, using the D-LSF GA with varying genetic operator frequencies. **b** Convergence rate of D-covariance GA. Frequency of mating (*black squares, solid line*), PM (*red circles, dotted line*) and OM (*blue triangles, dashed line*) are shown. Adapted from Ref. [17] with permission from John Wiley & Sons Ltd (Journal of Computational Chemistry), copyright 2012

calculations, along with the 95 % confidence levels, are available in Appendix B.2 for all genetic operators at differing frequencies.

The time consuming component of the GA search-algorithm is the FE stage. The average number of FEs, in GA searches converged to the GM, for D-LSF and D-covariance are very similar in value (Fig. 3.16a), with the 95 % confidence levels overlapping in all circumstances except at 0 %. This anomaly at 0 % can be associated with the limited number of converged results from which to extract this data; Fig. 3.15 shows convergence at 0 % PM was 2 % and 8 % for D-LSF and D-covariance, respectively. The average number of FEs increases for D-LSF and D-covariance with increasing frequency of the PM genetic operator, equalling an increase of 346 % and 110 %, respectively. For D-LSF the anomaly at 0 % PM can be held responsible for the over-exaggerated total percentage increase in total FEs; if this value is discounted then the increase in FEs for D-LSF from 10 % $\rightarrow$ 100 % is only 78 %. In all cases the relationship between FEs and increasing PM frequency is pseudo-linear; the same can also be said for mating and OM (see Appendix B.2).

We note that, with regards search efficiency, a two-fold improvement in convergence would be irrelevant if offset by an increase in the number of FEs by the same factor. The convergence can be evaluated alongside the FEs to give an overall value for the success of the search algorithm, when compared to performing an equivalent search in the absence of the genetic operator of interest, using the relationship:

$$\tau = \frac{C_i}{C_0} \cdot \frac{E_0}{E_i} \tag{3.16}$$

where C_i is the convergence rates (out of 100) with a genetic operator implemented at a prescribed frequency, compared to without the genetic operator (C_0). Likewise, E_i is the number of FE required with a genetic operator frequency compared to without the using the genetic operator (E_0). $\tau > 1$ corresponds to improved search efficiency, where as $\tau < 1$ corresponds to decreased efficiency compare to calculations in the absence of the genetic operator. In the case of the D-GA, the percentage improvement in convergence rate strongly outperforms the percentage increase in FEs, in all cases, making convergence the dominant measure of success as the frequency of genetic operator is increased; i.e. $\tau > 1$. A maximum for τ is seen with the mating operator set at 100 %, using the covariance FE, where $\tau = 10.3$ (Fig. 3.17). However this figure is somewhat skewed by the low convergence at 0 % frequency of the mating operator (1 %).

HEs remove the need to recalculate $\sigma_{\theta,\phi,\psi}$ for values of θ, ϕ and ψ that have already been visited, dramatically reducing FEs and thus calculation time. The average number of HEs in correctly converged searches for D-LSF, D-covariance, L-LSF and L-covariance GA searches are displayed in Fig. 3.16b. The average number of HEs conducted is greater than the number of FEs in all calculations. For D-type searches the gradient of change, again with respect to increasing PM frequency, is slightly negative above 20 % due to the increased exploration of new search areas. The low number of converged results for 20 % PM frequency and less make it

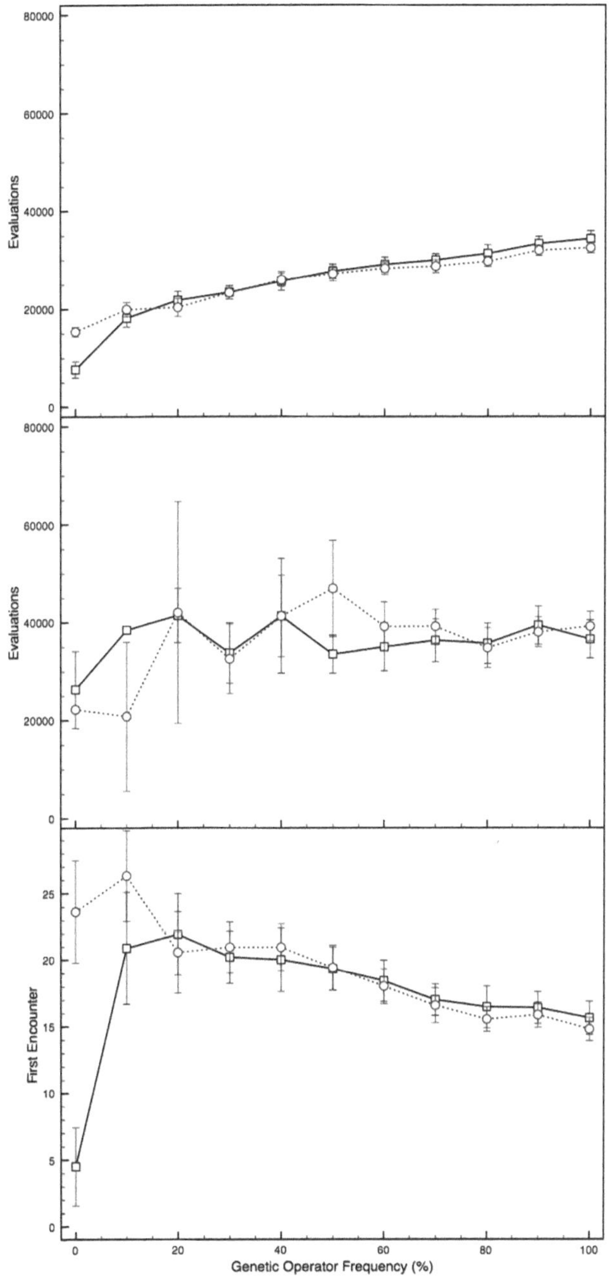

Fig. 3.16 For Ih structure: **a** The mean number of FEs required for converged GA runs with variation of PM (**b**). The mean number of HEs for converged GA runs. **c** The mean FENs for converged GA runs. The D-LSF (*black squares, solid line*) and D-covariance (*red circles, dotted line*) GAs are shown. Error bars are given at the 95 % confidence level. Adapted from Ref. [17] with permission from John Wiley & Sons Ltd (Journal of Computational Chemistry), copyright 2012

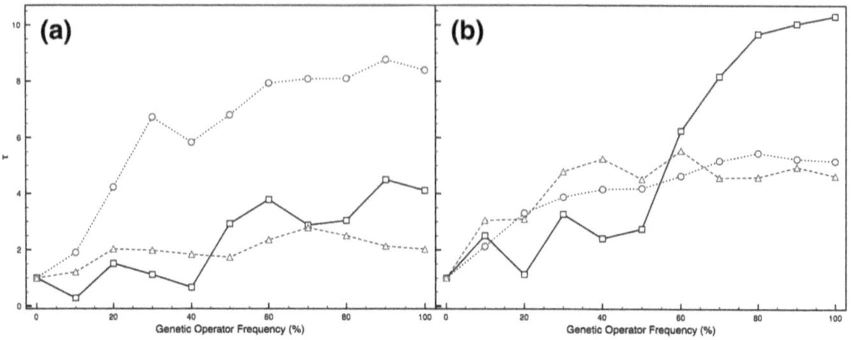

Fig. 3.17 For Ih structure: **a** τ for global optimisation searches using the D-LSF GA with varying genetic operator frequencies. **b** τ for global optimisation searches using the D-covariance GA. Frequency of mating (*black squares, solid line*), PM (*red circles, dotted line*) and OM (*blue triangles, dashed line*) are shown

difficult to confidently identify trends at the lower frequency levels (the large 95 % confidence levels, especially for D-covariance, illustrates the large variance in the data).

Figure 3.16c shows the mean FENs for converged D-LSF and D-covariance GA searches as a function of increasing PM frequency. From 20 % PM frequency upwards a negative gradient is seen for both D-LSF and D-covariance, implying improved success in the number of generations required by the GA to find a GM. For PM frequency below 20 % several anomalies are seen, especially D-LSF with 0 % PM which offers an unusually low value of 4.5. Similar features are visible for use of the mating operator in the D-LSF and D-covariance searches below 20 %. We can be confident in attributing these values to the sparsity of converged data, as previously mentioned for FEs and HEs, and this observation is supported by the relatively large 95 % confidence error bars displayed.

These observed trends hold true for the use of mating and OM genetic operators (see Appendix B.2 for data). The D-LSF GA with varied mating frequency shows increasing FEs, with slightly declining HEs and decreasing FENs above 10 % frequency. The limited data set may be responsible for the anomalies at low frequencies of the mating operator: convergence < 2 % for 10 % mating frequency and less. The OM operator shows a greater real value increase in FEs than mating and PM, with stable HEs and slightly decreasing mean FENs.

For D-covariance, FEs increase with mating frequency, and HEs and FENs decrease, showing similar trends to previous results above 20 % operator frequency; similar results are seen for changes in OM frequency. Values below this 20 % frequency are varied, implying that the converged dataset is too small for statistical analysis.

Decahedra

Convergence rates for a GA implemented search to find values of θ, ϕ and ψ, for a I-Dh STEM simulation to match a model solution are shown in Fig. 3.18. Varying frequencies of the genetic operators mating, PM and OM are examined. Data tables outlining convergence rates are available in Appendix B.2. Random-search calculations, using a random start-point and no minimisation, yielded results for the average number of FEs required to locate a GM (Table 3.5) where, as with the Ih random search, no minima were found. Hypothetically the ratio of searches to success should be significantly less than for the Ih, due to the reduced symmetry of the I-Dh structure (D_{5h}).

For the D-LSF and D-covariance GAs we see improved convergence with increasing frequency of the genetic operators, for mating and both mutation schemes (Fig. 3.18a, b). Convergence is low for mating at low frequencies, but rises rapidly as the frequency increases to 100 % (D-LSF: 79 %, D-covariance: 63 %). Of the mutation operators PM shows greater influence on convergence than OM; however, it must be added that both improve convergence as their frequency increases. Increasing PM frequency improves convergence from 27 to 91 % in the D-LSF GA, whilst for D-covariance a similar improvement of 14–87 % is seen (the peak convergence of 92 % for D-covariance is at 90 % PM frequency).

We notice here that some degenerate orientations for the I-Dh structure return slightly different values of $\sigma_{\theta,\phi,\psi}$, with a numerical difference of 10^{-6} %. Though slight, this has an effect on the convergence statistics and needs to be taken into consideration as degenerate orientations should, in theory, return equivalent $\sigma_{\theta,\phi,\psi}$ values. Updated convergence rates are plotted in Fig. 3.18c; we can see now that the D-covariance search is as competitive in terms of efficiency as the D-LSF, if not better, with a peak convergence of 97 %, for 100 % PM frequency.

All genetic operator frequency increases result in a greater number of FEs, however mating is far less responsive to this increase than mutation. The D-LSF and D-covariance GAs are very similar in the number of FEs computed, and the HEs are much greater in number than the FEs. Increasing PM frequency in D-LSF and D-covariance exhibit a noticeable downwards gradient in HEs.

Table 3.5 Benchmark tests using random searches (i.e. no GA) to locate the global minimum configuration for I-Dh. Calculations were performed without minimisation enabled, and run over a time period of 100 h each

Calculation type	FE type	FEs	Minimum found	Ratio
Point	LSF	1,726,012	0	–
	Covariance	1,671,001	0	–

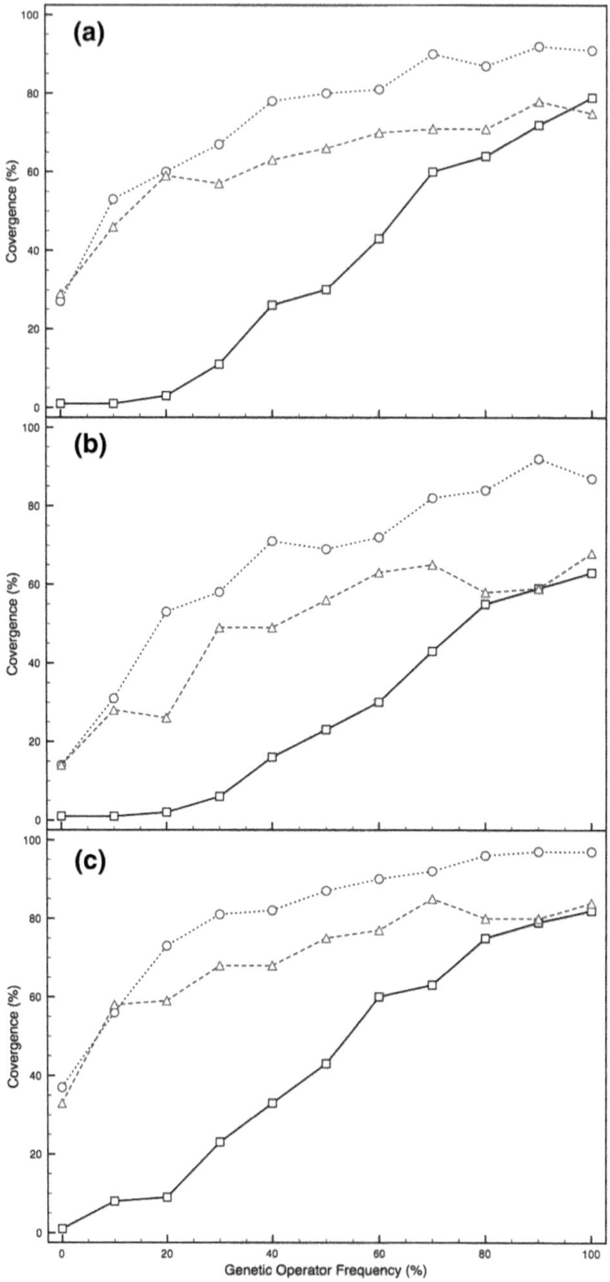

Fig. 3.18 For I-Dh structures: **a** Convergence rate of D-LSF GA, with varying genetic operator frequencies. **b** Convergence rate of D-covariance GA. **c** Convergence rate of D-covariance, corrected for degenerate structures. Key as for Fig. 3.15. Adapted from Ref. [17] with permission from John Wiley & Sons Ltd (Journal of Computational Chemistry), copyright 2012

3.6.2 Lamarckian Genetic Algorithm (L-GA)

Icosahedra

Convergence rates to the global minimum (GM) for L-GAs, searching for the global optimum values of θ, ϕ and ψ, are shown in Fig. 3.19, with varying frequencies of the mating and mutation (PM and OM) genetic operators [Data tables are available in Appendix B.3]. Random search calculations, using random start-points and no global minimisation method, yielded comparable results for the average number of FEs required to locate a minima for a variety of local minimisation techniques, as listed in Table 3.6. The ratio "FEs conducted : Minima found" from this data allows direct comparison to the efficiency of equivalent L-GA searches; in this instance the minimum ratio for random searches is 1 373.6:1 for a search conducted using Powell's method of local minimisation, coupled with a LSF-type FE, with all other ratios falling below 10,500:1. This is a considerable improvement compared to the non-locally minimised random searches, where no minima were found (Table 3.4).

L-GA searches result in high convergence even at low genetic operator frequencies, with increased frequency of genetic operators giving improved convergence in all L-GAs. For an L-LSF GA, using univariate minimisation, convergence at 0 % frequency of the genetic operators mating (73 %), PM (83 %) and OM (80 %) are considerably higher than seen for the D-GA (~70 %) even at maximum genetic operator frequency. This increases with operator frequency, with a maximum of 97 % for OM (Fig. 3.19a). This is improved upon by Powell's method of local minimisation, where convergence at 0 % genetic operator frequency is 84 %, 88 % and 87 % for mating, PM and OM, respectively. 100 % convergence is achieved when the OM operator is set to 100 %; with the mating (98 %) and PM (97 %) operators achieving similarly high maxima (Fig. 3.19b).

The multivariate L-LSF GA (Fig. 3.19c) gives improving convergence to the correct answer with increasing use of the genetic operators from 0 % → 100 % frequency: mating convergence (71 % → 88 %), PM (86 % → 94 %) and OM (73 % → 95 %); though, as with other L-GAs, the gradient of these improvements in convergence are less than for the D-GA due to success at low operator frequency.

Table 3.6 Benchmark tests using random, non-global, minimisation searches. Calculations were performed both with minimisation enabled, as shown in the column "Calculation Type", over a time period of 100 h each. "Ratio" corresponds to the number of FEs per minimum found

Calculation type	FE type	FEs	Minimum found	Ratio
Univariate	LSF	2,769,603	1,550	1,786.8
	Covariance	2,584,971	1,382	1,870.5
Powell	LSF	2,872,124	2,091	1,373.6
	Covariance	2,191,125	1,511	1,450.1
Multivariate	LSF	7,831,093	784	9,988.6
	Covariance	7,806,088	763	10,230.8

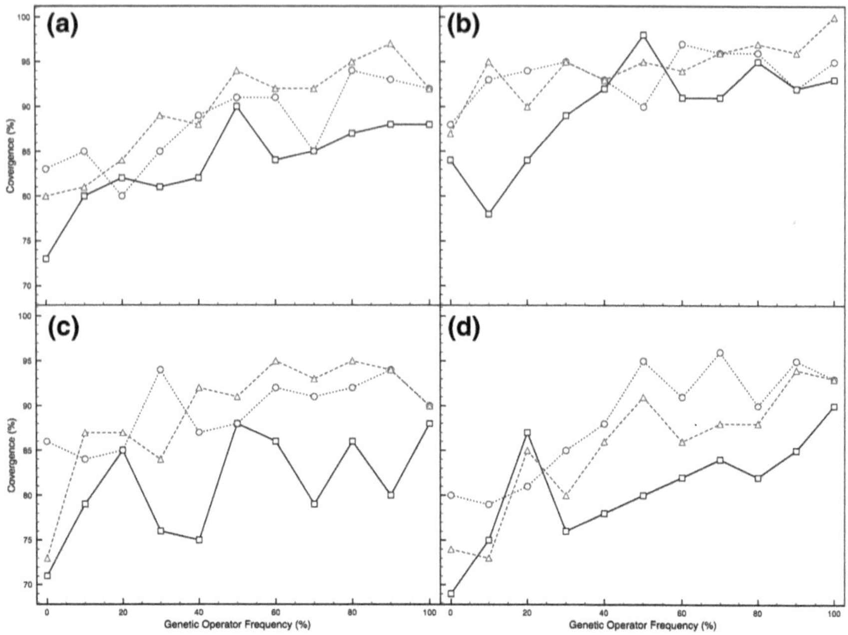

Fig. 3.19 For the Ih structure: **a** Convergence rate for L-LSF GA with univariate local minimisation. **b** Convergence rate for L-LSF GA with Powell's method of local minimisation. **c** Convergence rate for L-LSF GA with multivariate local minimisation. **d** Convergence rate for L-covariance GA with multivariate local minimisation. Frequency of mating (*black squares, solid line*), PM (*red circles, dotted line*) and OM (*blue triangles, dashed line*) are shown. Adapted from Ref. [17] with permission from John Wiley & Sons Ltd (Journal of Computational Chemistry), copyright 2012

The same observation is seen for the L-covariance GA (Fig. 3.19d), tested with the multivariate local minimisation method: mating (69 % $\rightarrow$ 90 %) and mutation (PM: 79 % $\rightarrow$ 96 %, OM: 74 % $\rightarrow$ 94 %) with increasing operator frequency; the increase in success with a higher frequency of genetic operator is not as significant as seen for the D-covariance GA. The limited improvement in convergence for L-GA, when compared to the D-GA, can be associated with the relatively high convergence achieved by implementing the local minimisation of each population member between generations. This observation can be consolidated by the successful results of the random local-minimisation search (Table 3.6) compared to the non-minimising random search (Table 3.4). In all the L-GAs tested the frequency of mating appears to have slightly less influence than mutation on convergence, similar to observations for the D-GAs.

Data tables, outlining minimum, average and maximum number of FEs, HEs and FENs for converged calculations, along with the 95 % confidence levels, are available in Appendix B.3 for all genetic operators at differing frequencies, and plotted for PM in Fig. 3.20. L-LSF searches using the univariate and Powell's method local minimisation methods return a low average number of FEs, with comparison to the

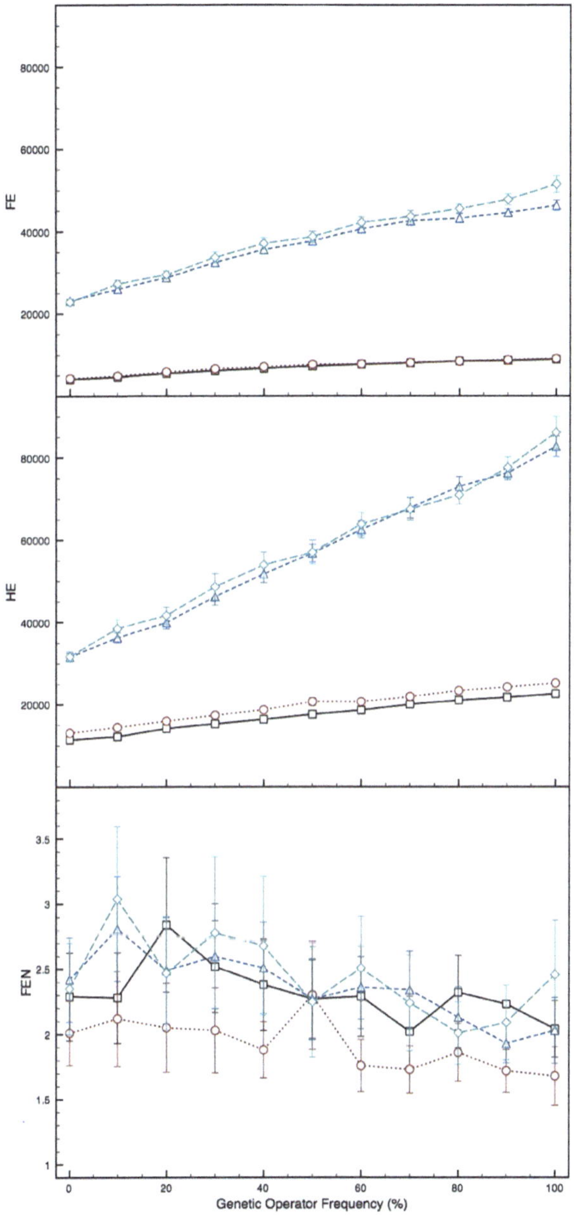

Fig. 3.20 For Ih structure: **a** The mean number of FEs required for converged L-GA runs with variation of PM. **b** The mean number of HEs for converged L-GA runs. **c** The mean FENs for converged L-GA runs. The L-LSF searches using univariate (*black squares, solid line*), Powell's method (*red circles, dotted line*) and multivariate (*blue triangles, short-dashed line*), and multivariate L-covariance (*green diamonds, long-dashed line*), GAs are shown. Error bars are given at the 95 % confidence level. Adapted from Ref. [17] with permission from John Wiley & Sons Ltd (Journal of Computational Chemistry), copyright 2012

multivariate L-LSF and L-covariance (Fig. 3.20a). The mean number of FEs increases with operator frequency: multivariate L-GA searches follow a similar upwards gradient to the D-GA, though there is some divergence between L-covariance and L-LSF as the PM operator frequency approaches 100 %, with the L-covariance GA requiring slightly more FEs (At 100 % PM, L-LSF: 46332 ± 1217, L-covariance: 51509 ± 2023). Expectedly, the rate of change for the univariate and Powell's method are much lower than for the multivariate method as they are less computationally expensive. This, coupled with the high convergence of these methods, is a positive result; though the mean number of FEs for a minimum is still lowest using the random locally-minimising search.

In general, the difference between the number of FEs required for D-GA and the multivariate locally minimised L-GA searches (ΔFE) is in the range $5{,}000 < \Delta\text{FE} < 15{,}000$, equating to an increase of up to 50 % in the number of FEs required by the L-GA compared to the D-GA; however for the univariate and Powell's local minimisation the mean number of FEs is 50 % lower than for the D-GA. In all cases, the relationship between FEs and increasing PM frequency is pseudo-linear; the same can also be said for mating and OM (see Appendix B.3).

The average number of HEs in correctly converged searches for L-LSF and L-covariance GA searches are displayed in Fig. 3.20 b. As with the D-GA searches, the average number of HEs conducted is greater than the number of FEs in all calculations.

For multivariate-minimising searches, we can see the average number of HEs for LSF and covariance are very similar, with the 95 % confidence levels overlapping in all cases. For univariate and Powell's method, the mean number of HEs is considerably less than multivariate. Error bars on these latter methods are also small (<300 HEs) compared to the multivariate local minimisation. The positive gradient of the HEs with respect to increased PM frequency is greater than for FEs, in all cases, illustrating that HEs become more frequently used as we increase genetic operations, especially with the multivariate local minimisation method. This is in contrast to results for D-GAs where HEs decrease with increasing PM operator frequency, and intuitively implies that the search space is revisited during the local minimisation routines.

Figure 3.20c shows the mean FENs for converged L-LSF and L-covariance GA searches as a function of increasing PM frequency. We see that for the L-GA searches the mean FEN < 3 in all but one case (L-covariance, 10 % PM, FEN: 3.04) and the 95 % confidence levels are small (<0.6), much lower than seen for the D-GA. The lowest values are seen for Powell's method of local minimisation, with a minimum at 1.68 (100 % PM frequency). An anomaly is seen at 50 % PM frequency for the Powell's method, but the large error bars give indication that this is a statistical anomaly which could be corrected by repeating calculations over a larger number of runs e.g. 1,000.

All observed trends hold true for the use of mating and OM genetic operators (see Appendix B.3 for data), using all local minimisation methods. Increasing OM frequency gives increasing FEs and HEs, much higher than for the D-type searches, and a marginally decreasing mean FEN. Increased use of the mating operator also leads to increased FEs and HEs, and the mean FENs show a strong decreasing trend.

For the L-covariance GA, using multivariate local minimisation, we see identical trends as for the multivariate L-LSF, with increasing use of the mating and OM operators resulting in increasing FEs and HEs, with OM having a significantly greater effect on the quantity, whilst mean FENs decrease more for mating ($3.7 \rightarrow 2.29$) than OM ($3.08 \rightarrow 2.02$).

Decahedra

Convergence rates for a L-GA implemented search to find optimised values of θ, ϕ and ψ, for a I-Dh STEM simulation, matched to a model solution, are shown in Fig. 3.21. Varying frequencies of the genetic operators mating, PM and OM are examined. Data tables outlining convergence rates are available in Appendix B.3. Random-search calculations, using a random start-point and appropriate local-minimisation, yielded comparable results for the average number of FEs required to locate a

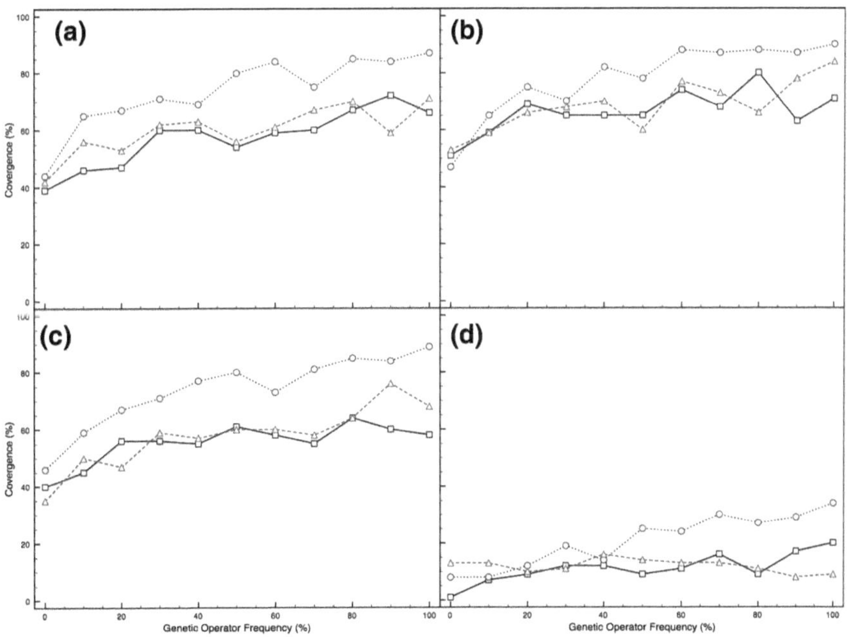

Fig. 3.21 For I-Dh structures: **a** Convergence rate for L-LSF GA using univariate minimisation. **b** Convergence rate for L-LSF GA using Powell's method local minimisation. **c** Convergence rate for L-LSF GA using multivariate minimisation. **d** Convergence rate for L-covariance GA. Key as for Fig. 3.19. Adapted from Ref. [17] with permission from John Wiley & Sons Ltd (Journal of Computational Chemistry), copyright 2012

GM (Table 3.7). The minimum ratio is 4,976.4:1, for the Powell's method locally-minimised covariance-type random search, with the highest ratio being 44,047.0:1 for the multivariate locally-minimised search. The ratio of searches to success is significantly less than for the Ih, due to the reduced symmetry of the I-Dh structure (D_{5h}).

Using univariate local minimisation routines, convergence improvements range over 29 – 43 % with increasing genetic operator frequencies (Fig. 3.21a), again with PM (44 % $\rightarrow$ 87 %) having a greater influence than mating (39 % $\rightarrow$ 72 %) and OM (42 % $\rightarrow$ 71 %). For Powell's local minimisation routine coupled into the L-LSF GA, these values improved for mating (51 % $\rightarrow$ 80 %), PM (47 % $\rightarrow$ 90 %) and OM (53 % $\rightarrow$ 84 %) (Fig. 3.21b). For multivariate local minimisation, the L-GA regime yields relatively small improvements with the increased frequency of genetic operators. Increased convergence is seen for the L-LSF GA (Fig. 3.21c) with increasing frequency of mating (40 % $\rightarrow$ 64 %), OM (35 % $\rightarrow$ 76 %) and PM (46 % $\rightarrow$ 89 %) operators. None of these convergence rates match the maximum seen for the D-GA searches; nor do they match the results achieved by Powell's local minimisation method.

The multivariate L-covariance GA (Fig. 3.21d) yields poor convergence results, illustrating a potentially frustrated function landscape. Mating and PM are shown to improve convergence rates where as OM yields no improvement, with convergence in the range of 9–16 %. Examination of other competitive local minima found for the L-covariance GA quickly identifies degenerate orientations which the GA converges to more frequently (Fig. 3.22), perhaps as a result of a minor numerical inconsistency in the original coordinate file: the difference between the covariance FE for these two orientations is less than 10^{-6} %. With this taken into consideration the overall convergence for the L-covariance GA ranges between 36 % $\rightarrow$ 75 %, 51 % $\rightarrow$ 77 % and 59 % $\rightarrow$ 78 % for mating, PM and OM, respectively, and is comparable to the search success of the multivariate L-LSF GA.

All increases in genetic operator frequencies result in a greater number of FEs for all local minimisation methods, however the quantity of FEs is far less responsive to mating than mutation. For both D-type and multivariate L-type searches the LSF and

Table 3.7 Benchmark tests using random searches (i.e. no GA) to locate the global minimum configuration for I-Dh. Calculations were performed with local minimisation enabled, as shown in the column "Calculation Type", and run over a time period of 100 h each. "Ratio" corresponds to the number of FEs per minimum found

Calculation type	FE type	FEs	Minimum found	Ratio
Univariate	LSF	2,084,985	176	11,846.5
	Covariance	2,735,042	365	7,493.3
Powell	LSF	2,749,858	450	6,110.8
	Covariance	2,836,568	570	4,976.4
Multivariate	LSF	7,884,420	179	44,047.0
	Covariance	7,834,739	250	31,339.0

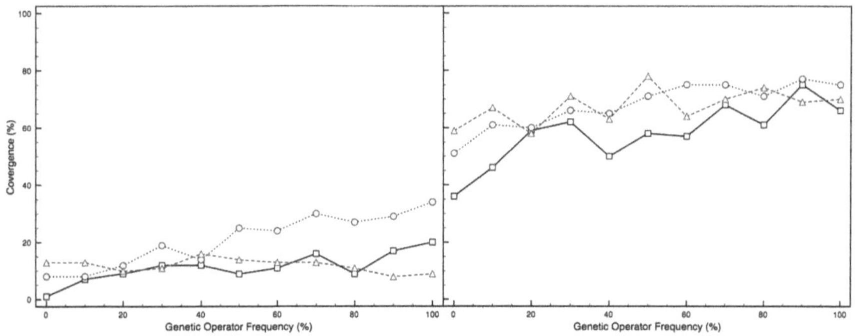

Fig. 3.22 Comparison of competitive minima for L-covariance searches on the I-Dh structure. *Left*: Convergence rate to true minimum result. *Right*: Convergence rate to all degenerate minima, not taking into consideration small (10^{-6} %) differences in $\sigma_{\theta,\phi,\psi}$. Key as for Fig. 3.19

covariance are very similar in the number of FEs computed; the number of FEs are much less for univariate and Powell's method of local minimisation. The quantity of HEs is much greater in number than the FEs, throughout all minimisation methods.

For L-GAs searching the I-Dh landscape the mean FENs fail to significantly decrease with increasing genetic operator frequency, similar to the results seen for L-GAs searching the Ih landscape. As an example, increasing PM frequency in D-LSF and D-covariance exhibit a noticeable downwards gradient, whilst for multivariate L-LSF and L-covariance the same is not true.

We can use τ (Eq. 3.16) to give a measure of the relative improvement of GA efficiency with increasing genetic operator frequency: for the L-GAs, $\tau < 1.5$ in all cases, and always decreasing in value with increasing frequency of the genetic operator beyond 30 % (Fig. 3.23). This illustrates there is little significant computational gain in increasing the frequency of the genetic operators beyond 30 %, other than the reassurance that the GM is found.

3.6.3 *Tolerance to Noise*

Convergence rates for the D-GA and L-GA were tested against images with increasing levels of controlled background noise, defined as a function of the maximum intensity on the original HAADF-STEM image. Convergence rates prove consistent up to 4 % noise levels, after which it was found that the correct orientation was no longer found as the GM by the search algorithms (Fig. 3.24). Convergence to other orientations increases significantly once noise levels are beyond the 4 % threshold identified. Therefore, we can be confident of the success of our genetic algorithm at low noise levels as a result of these measurements, justifying its suitability for experimental application.

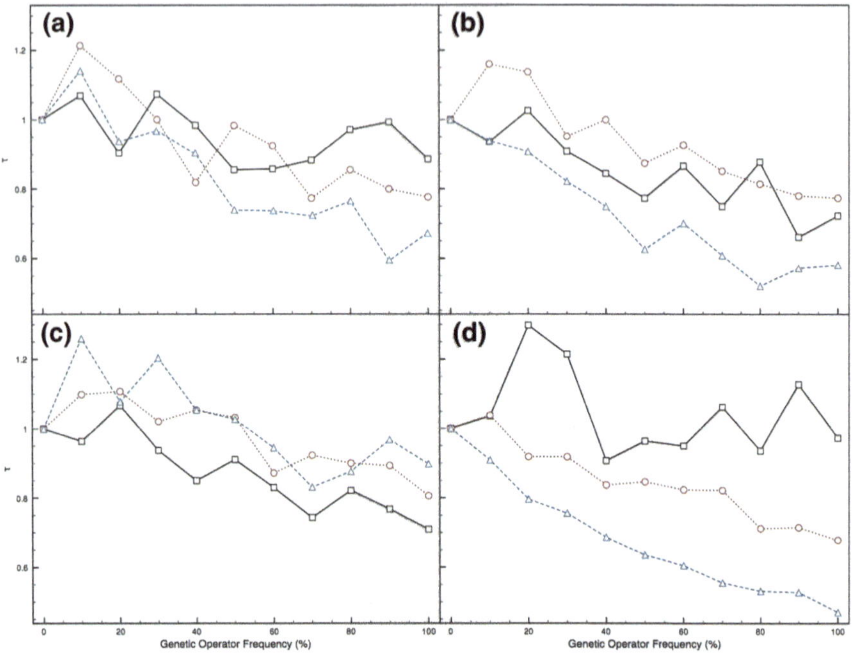

Fig. 3.23 For I-Dh structures: **a** τ for L-LSF GA using univariate minimisation. **b** τ for L-LSF GA using Powell's method local minimisation. **c** τ for L-LSF GA using multivariate minimisation. **d** τ for L-covariance GA, taking into consideration degenerate orientations with dissimilar $\sigma_{\theta,\phi,\psi}$ values. Key as for Fig. 3.19, with frequency of mating (*black squares, solid line*), PM (*red circles, dotted line*) and OM (*blue triangles, dashed line*) are shown. The "break-even" point is highlighted at $\tau = 1$

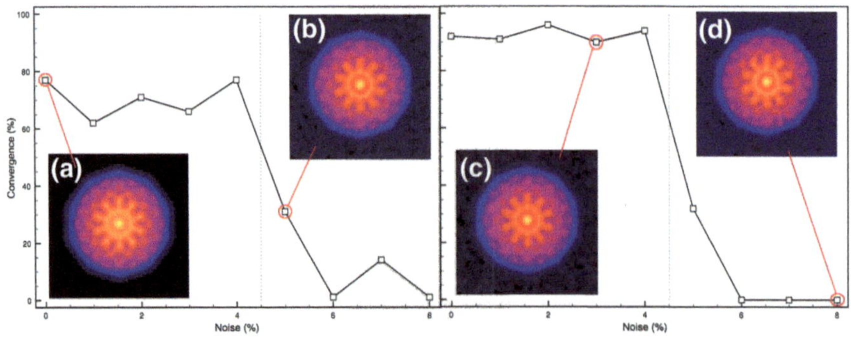

Fig. 3.24 Convergence rate of the GA using model solutions with additional noise to the correct solution. *Left* D-LSF GA search. *Right* L-LSF GA search, using multivariate local minimisation. *Inset* STEM images, in false colour: **a** 0 % noise; **b** 5 % noise; **c** 3 % noise; **d** 8 % noise. The *grey vertical line* at 4.5 % represents the transition between where the GA searches find the correct orientation as the GM (for 4 % and below), and where other orientations other than the correct orientation return better $\sigma_{\theta,\phi,\psi}$ values (5 % and above)

3.7 Conclusions

We have discussed and developed the implementation of a HAADF-STEM model, coupling it successfully with a genetic algorithm (GA). A model was proposed, taken from the work of Curley et al. [16], and developed to take into consideration (a) the effect of an elemental species atomic number and (b) the radius of an elemental species, as defined by two constant parameters. Furthermore, we implemented statistical methods to compare modelled HAADF-STEM images with experimental, or other modelled, images. Of the comparative methods implemented, LSF shows greater susceptibility to the configuration of an experimental image: data must be centralised, with intensities normalised, otherwise results cannot be considered valid. Covariance offers an alternative comparative method, and this has also been implemented and tested. A residual ($\sigma_{\theta,\phi,\psi}$) has been used to gauge the relative suitability of a theoretical structure's orientation, dependent on the Tait-Bryan angles θ, ϕ and ψ, to alternative orientations, when compared to a predefined solution.

Application of our HAADF-STEM model and numerical comparative methods was tested initially on monolayer-protected Au_{38} clusters, with a linear search conducted on 5° steps. From this, we were able to corroborate our results with those of Wang et al. who identify a prolate structure from their experimental results, previously suggested by Pei et al. and Lopez-Avecedo et al [21, 22]. Of our results, the LSF proves more consistent at producing results that qualitatively match the experimental image we have used; however this is, of course, subjective. Subtraction of noise from the experimental results proves ineffective at helping the characterisation process, predominantly through the removal of valid features as well as the spurious substrate-derived intensities which we wanted to remove.

Method improvements were identified, of which some have been addressed. Though thorough, the linear search method, with 5° steps, is inefficient at exploring the entire function hypersurface adequately, with a coarse search grid necessary to ease computation. Local and global minimisation was therefore implemented: local minimisation has been tested with a variety of regimes including univariate, Powell's method and multivariate minimisation. A genetic algorithm was implemented as a global search algorithm, and tested using 1° steps.

Through parameterisation we were able to customise the GA such that it operates as efficiently as possible. We show, using an Icosahedron (Ih) and Decahedron (I-Dh) as test subjects, that the increasing frequency of the genetic operators within the GA (crossover and mutation) improve convergence rates to the GM. Mutation of the population parents proves most effective at improving convergence rates, whilst mating often proves to be the least effective. However, the frequency of mating also has the least effect on the number of FEs.

D-GAs prove to be efficient at finding an optimum orientation of a test cluster, when compared to a random searching algorithm, however the implementation of a computationally non-intensive local minimisation method significantly improves this: use of either univariate or Powell's method for local minimisation, coupled with a GA to form a Lamarckian-GA, prove more efficient than a D-GA in regards finding the optimum nanocluster orientation without requiring a large amount of computational resources. The same cannot be said for the multivariate L-GA, despite its relatively high convergence rates: Fig. 3.25 compares the number of FEs required for the multivariate L-LSF GA to the D-LSF GA. The spread of FEs (maximum – minimum) is smaller for the D-LSF than the L-LSF counterpart, leading to small 95 % confidence level bars; clearly the multivariate L-GA requires considerably more FEs overall.

For D-GAs the quantity of HEs is found not to significantly increase with genetic operator frequency, however for L-GAs HEs are always greater than FEs; the implication from this is search space is revisited, most frequently during the local minimisation process. Also, the mean first encounter (FEN) of the GM decreases as the genetic operator increases for the D-GA, whereas or L-GAs this decrease is less significant as the GM search is considerably improved by the downhill-gradient driven local minimisation process. Of the structures used for benchmarking, overall convergence proved more successful to the Ih structure, due to the increased symmetry compared to the I-Dh. This is something that can be utilised in future for symmetric structures, where search space can be restricted to just the non-repeating orientations, vastly improving search speed and the efficiency of the genetic operators.

We have also tested the resilience of our GA search methods to increasing background noise levels, as a function of maximum intensity. A threshold was found at 4 %, after which convergence to the correct orientation diminishes.

Clearly, whilst the Darwinian-GA offers an improvement on randomly searching the 3D hypersurface without minimisation, random locally minimising searches prove more efficient than Lamarckian-GAs by an order of magnitude. The most efficient local minimisation method proves to be Powell's method, where an additional search vector improves the analytical minimisation process; in some circumstances this may be because the search vector allows "tunnelling" through an otherwise impassable function barrier. Further work is needed to check if the problems we have used for benchmarking are simply too easy, and to see if alternative methods of global optimisation prove more suitable.

In addition to the implemented search algorithms, the power of parallel computing has been efficiently harnessed such that both shared memory (OMP) and non-shared memory (MPI) parallelisation has been installed into our search algorithm with success. Whilst not significantly tested in this work, the possibility of using MPI to search across a database of experimental images improves the potential use of this search method in further work by experimental collaborators.

Deficiencies in our model cannot be ignored, and should be considered for attention in further work. In smaller nanoclusters, the mobility of atoms and molecules under the electron beam makes it difficult to easily identify consistent morphologies, with system "snapshots" being the most achievable results. This can perhaps be addressed by the use of either (a) Molecular Dynamics (MD) to simulate the evolution of clusters at temperature, through the calculation of the Debye-Waller factor, or (b) using stochastic methods to restructure theoretical clusters so that they better fit our experimental images. At an early stage the former could be approximated using coordinate number: though this would need initial calibration it would restrict the need for repeated MD simulations. With the advent of modern technology,

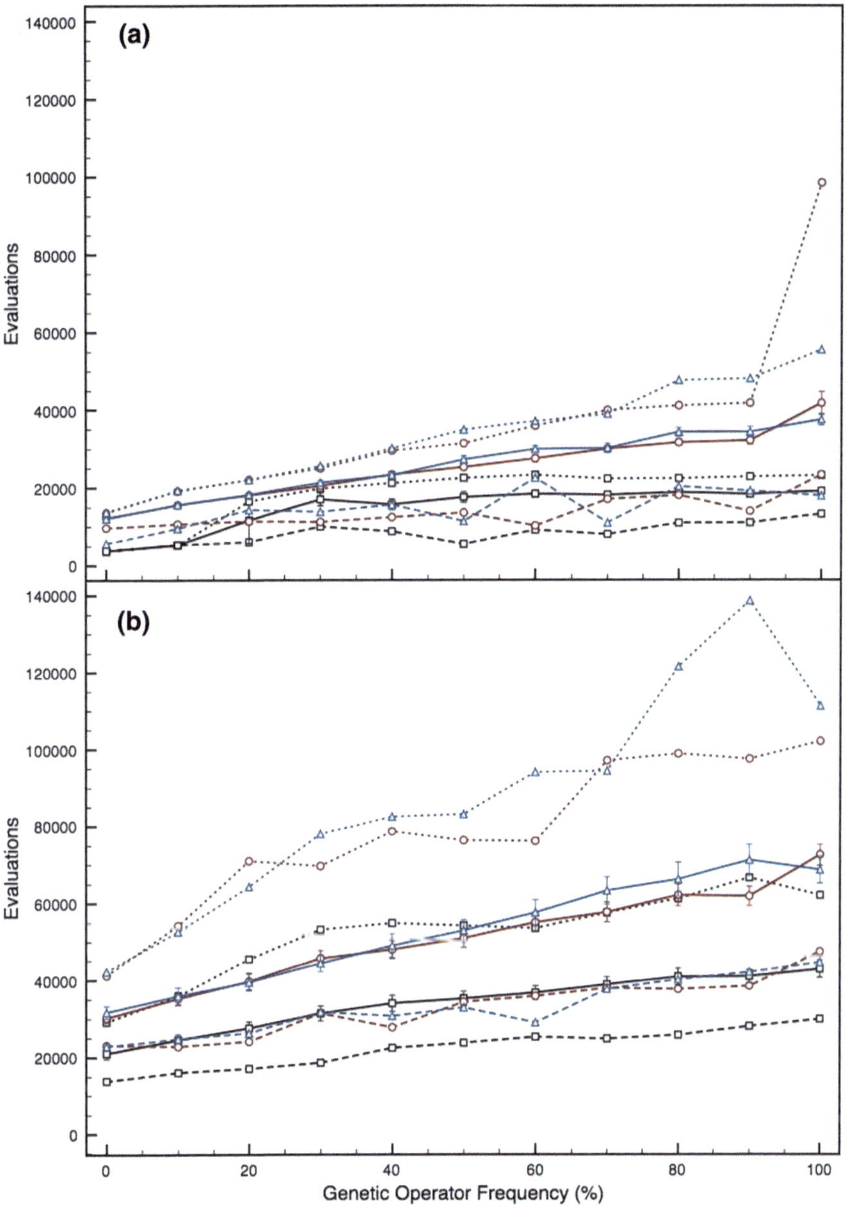

Fig. 3.25 For I-Dh structures: **a** Average number of FEs for D-LSF GA. **b** Average number of FEs for L-LSF GA. The average number of FEs (*solid lines*) are given with 95 % confidence levels, along with minimum (*dashed lines*) and maximum (*dotted lines*) values. The different genetic operations of mating (*black squares*), MP (*red circles*) and MO (*blue triangles*) are plotted. Adapted from Ref. [17] with permission from John Wiley & Sons Ltd (Journal of Computational Chemistry), copyright 2012

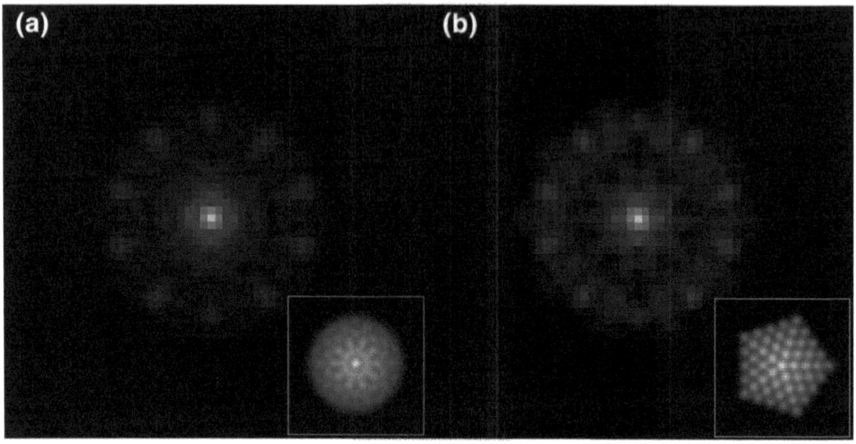

Fig. 3.26 Fast Fourier Transforms of HAADF-STEM modelled images: **a** Icosahedron and **b** decahedron. Original image is given as an *inset*

perhaps low-temperature experimental work will allow for more thermally stable cluster analysis in the future, aiding our work; the inherent $3N - 6$ atomic degrees of freedom is a serious complication to overcome in all forms of structural determination work.

Improvements could also be made on the probe, currently represented by a Gaussian function but perhaps better represented using an Airy function. Also, for larger clusters, more efficient use of Fast Fourier Transforms (FFTs) could be used to compare crystal structures with experimental images as part of an initial structure search. FFTs are centralised, and thus the need for calculating the centre of mass for an image would also be removed; they also offer easy characterisation of crystalline structures, and would be suitable for structure pre-screening (Fig. 3.26). FFTs have been implemented in our code, but not fully tested to ensure suitability.

References

1. Z.Y. Li, N.P. Young, M. Di Vece, S. Palomba, R.E. Palmer, A.L. Bleloch, B.C. Curley, R.L. Johnston, J. Jiang, J. Yuan, Three-dimensional atomic-scale structure of size-selected gold nanoclusters. Nature **451**(7174), 46–48 (2008)
2. P.M. Voyles, D.A. Muller, J.L. Grazul, P.H. Citrin, H.-J.L. Gossmann, Atomic-scale imaging of individual dopant atoms and clusters in highly n-type bulk Si. Nature **416**(6883), 826–829 (2002)
3. P.E. Batson, N. Dellby, L.O. Krivanek, Sub-angstrom resolution using aberration corrected electron optics. Nature **418**, 617–620 (2002)
4. S. Van Aert, K.J. Batenburg, M.D. Rossell, R. Erni, G. Van Tendeloo, Three-dimensional atomic imaging of crystalline nanoparticles. Nature **470**, 374–377 (2009)

5. S. Bals, M. Casavola, M.A. van Huis, S. Van Aert, K.J. Batenburg, G. Van Tendeloo, D. Vanmaekelbergh, Three-dimensional atomic imaging of colloidal core-shell nanocrystals. Nano Lett. **11**, 3420–3424 (2011)
6. K. Watanabe, T. Yamazaki, I. Hashimoto, M. Shiojiri, Atomic-resolution annular dark-field stem image calculations. Phys. Rev. B **64**(11), 115432–115437 (2001)
7. E.J. Kirkland, *Advanced Computing in Electron Microscopy* (Plenum Press, New York/London, 1998)
8. D. Ferrer, D.A. Blom, L.F. Allard, S. Mejia, E. Perez-Tijerina, M. José-Yacamán, Atomic structure of three-layer Au/Pd nanoparticles revealed by aberration-corrected scanning transmission electron microscopy. J. Mater. Chem. **18**, 2442–2446 (2008)
9. F.L. Deepak, A. Mayoral, A.J. Steveson, S. Mejia-Rosales, D.A. Blom, M. José-Yacamán, Insights into the capping and structure of MoS_2 nanotubes as revealed by aberration-corrected STEM. Nanoscale **2**, 2286–2293 (2010)
10. K. Ishizuka, A practical approach for stem image simulation based on the fft multislice method. Ultramicroscopy **90**(2–3), 71–83 (2002)
11. A. Colli, E. Carlino, E. Pelucchi, V. Grillo, A. Franciosi, Local interface composition and native stacking fault density in ZnSe/GaAs(001) heterostructures. J. Appl. Phys. **96**(5), 2592–2602 (2004)
12. F. Cleri, V. Rosato, Tight-binding potentials for transition metals and alloys. Phys. Rev. B **48**(1), 22–33 (1993)
13. R. L. Johnston, Evolving better nanoparticles: genetic algorithms for optimising cluster geometries. Dalton Trans. 4193 (2003)
14. Z. Wang, O. Toikkanen, B.M. Quinn, R.E. Palmer, Real-space observation of prolate monolayer-protected Au_{38} clusters using aberration-corrected scanning transmission electron microscopy. Small **7**(11), 1542–1545 (2011)
15. K.A. Mkhoyan, S.E. Maccagnano-Zacher, E.J. Kirkland, J. Silcox, Effects of amorphous layers on ADF-STEM imaging. Ultramicroscopy **108**(8), 791–803 (2008)
16. B.C. Curley, R.L. Johnston, N.P. Young, Z.Y. Li, M. Di Vece, R.E. Palmer, A.L. Bleloch, Combining theory and experiment to characterize the atomic structures of surface-deposited Au_{309} clusters. J. Phys. Chem. C **111**(48), 17846–17851 (2007)
17. A.J. Logsdail, Z.Y. Li, R.L. Johnston, Development and optimization of a novel genetic algorithm for identifying nanoclusters from scanning transmission electron microscopy images. J. Comp. Chem. **33**(4), 391–400 (2012)
18. R.E. Roberson, R. Schwertassek, *Dynamics of Multibody Systems* (Springer, Berlin/New York, 1988)
19. V. Lepetit, P. Fua, *Monocular Model-Based 3D Tracking of Rigid Objects* (Now Publishers Inc., Delft, 2005)
20. H. Häkkinen, M. Walter, H. Grönbeck, Divide and protect: Capping gold nanoclusters with molecular gold-thiolate rings. J. Phys. Chem. B **110**(20), 9927–9931 (2006)
21. Y. Pei, Y. Gao, X.C. Zeng, Structural prediction of thiolate-protected Au_{38}: a face-fused bi-icosahedral Au core. J. Am. Chem. Soc. **130**(25), 7830–7832 (2008)
22. O. Lopez-Acevedo, H. Tsunoyama, T. Tsukuda, H. Häkkinen, C. M. Aikens, Chirality and electronic structure of the thiolate-protected Au_{38} nanocluster. J. Am. Chem. Soc. **132**(23), 8210–8218 (2010)
23. M. Walter, J. Akola, O. Lopez-Acevedo, P.D. Jadzinsky, G. Calero, C.J. Ackerson, R.L. Whetten, H. Grönbeck, H. Häkkinen, A unified view of ligand-protected gold clusters as superatom complexes. Proc. Natl. Acad. Sci. **105**, 9157–9162 (2008)
24. Z. Wang, Personal, Communication (2009)
25. J.C. Slater, Atomic radii in crystals. J. Chem. Phys. **41**(10), 3199 (1964)
26. D.E. Goldberg, *Genetic Algorithms in Search Optimization and Machine Learning* (Addison-Wesley, Reading, 1989)
27. M.T. Oakley, D.J. Wales, R.L. Johnston, Energy landscape and global optimization for a frustrated model protein. J. Phys. Chem. B **115**(39), 11525–11529 (2011)
28. C. Darwin, *On the Origin of Species* (John Murray, London, 1859)

29. R.L. Johnston, *Applications of Evolutionary Computation in Chemistry* (Springer, Heidelberg, 2004)
30. D.E. Clark, *Evolutionary Algorithms in Molecular Design* (Wiley-WCH, Weinheim, 2000)
31. J. Forshed, I. Schuppe-Koistinen, S.P. Jacobsson, Peak alignment of NMR signals by means of a genetic algorithm. Anal. Chim. Acta **487**(2), 189–199 (2003)
32. K.D.M. Harris, R.L. Johnston, B.M. Kariuki, The genetic algorithm: foundations and applications in structure solution from powder diffraction data. Acta Cryst. **54**(5), 632–645 (1998)
33. R. Leardi, Genetic algorithms in chemometrics and chemistry: a review. J. Chemometr. **15**(7), 559–569 (2001)
34. G. Jones, P. Willett, R.C. Glen, A.R. Leach, R. Taylor, Development and validation of a genetic algorithm for flexible docking. J. Mol. Biol. **267**(3), 727–748 (1997)
35. H.M. Cartwright, *Applications of Artificial Intelligence in Chemistry* (Oxford University Press, Oxford, 1993)
36. A.J. Logsdail, L.O. Paz-Borbon, R.L. Johnston, Structures and stabilities of platinum-gold nanoclusters. J. Comput. Theor. Nanosci. **6**, 857–866 (2009)
37. R. Ismail, R.L. Johnston, Investigation of the structures and chemical ordering of small pd-au clusters as a function of composition and potential parameterisation. Phys. Chem. Chem. Phys. **12**(30), 8607–8619 (2010)
38. L.O. Paz-Borbón, R.L. Johnston, G. Barcaro, A. Fortunelli, Structural motifs, mixing, and segregation effects in 38-atom binary clusters. J. Chem. Phys. **128**(13), 134517 (2008)
39. T. Dandekar, P. Argos, Potential of genetic algorithms in protein folding and protein engineering simulations. Protein Eng. **5**(7), 637–645 (1992)
40. G.A. Cox, T.V. Mortimer-Jones, R.P. Taylor, R.L. Johnston, Development and optimisation of a novel genetic algorithm for studying model protein folding. Theor. Chem. Acc. **112**(24), 163–178 (2004)
41. W.H. Press, S.A. Teukolsky, W.T. Vetterling, B.P. Flannery, *Numerical Recipes in C* (Cambridge University Press, Cambridge, 1997)
42. J.B.P.A. de Lamarck, *Zoological Philosophy: An Exposition with Regard to Natural History of Animals* (Hafner Pub Co., New York, 1963)
43. F. Glover, F. Laguna, *Tabu Search* (Kluwer Academic Publishers, Norwell, 1997)
44. R.P. Gupta, Lattice relaxation at a metal surface. Phys. Rev. B **23**(12), 6265–6270 (1981)
45. C. Kittel, *Introduction To Solid State Physics* 6th edn. (Wiley, New York, 1986)
46. http://www.bear.bham.ac.uk/bluebear/

Chapter 4 A First-Principles Study of the Soft-Landing of Au_{16} on Graphite

4.1 Introduction

Recent efforts in cluster science have focused on phenomena associated with nanocluster-substrate interactions [1]. The atomic smoothness and chemical inertness of highly ordered pyrolytic graphite (HOPG) means that it is often used to support nanoparticles, for instance alkali [2] and noble [3, 4] metal clusters. However, noble metal atoms and clusters are very weakly adsorbed on HOPG (and its derivatives such as carbon nanotubes, CNTs) and diffuse easily which leads to formation of larger aggregates (sintering) [5, 6]. Recently, limiting diffusion has been approached by the introduction of surface defects [5, 7]. One effective method for preventing diffusion after deposition is provided if clusters impact on the substrate with enough kinetic energy to distort the local lattice, or displace an atom from the surface [5, 8], thus introducing a defect. Alternatively, surfaces can be pre-treated to introduce defects [9], on to which clusters can be soft-landed. Recent experimental observations have shown Au atoms prefer to adsorb in the plane with graphene [10], and this has been confirmed by first-principle calculations [11], whilst in the absence of defects it has been shown that inducing strain in an HOPG surface aids adsorption of small gold clusters [12].

Au nanoclusters are of continued scientific interest due to their size-dependent properties [13, 14] and potential applications in catalysis [15–17]. Unlike monovalent alkali metals, small uncharged gold clusters are shown to favour planar structures for $N < 12$ [18–20], with this trend being extended to $N = 12$ for anionic clusters [21]. Au_{20} forms a pyramidal structure with T_d symmetry; the large highest occupied molecular orbital (HOMO) to lowest unoccupied molecular orbital (LUMO) band gap (1.8 eV) emphasises its stability [22, 23]. Removal of the four vertex atoms from Au_{20} followed by structural relaxation leads one to a hollow cage-like structure for the Au_{16} cluster [24], with T_d-symmetry maintained. This cage-like structure has been identified experimentally for anionic Au_n^- ($n = 14$–18) [19, 24], generating interest due to the possibility of endohedral doping of the central void with alternative elements [25–27]. For the neutral Au_{16} particle the cage-like structure is shown to be

A. J. Logsdail, *Computational Characterisation of Gold Nanocluster Structures*, Springer Theses, DOI: 10.1007/978-3-319-01493-7_4,

a local, but not global, minimum in recent calculations by Chen et al. [28]. They also show that the cage-like structure is maintained upon electron detachment from Au_{16}^- (the global minimum in this case), leading to conclusions that additional assistance is necessary to overcome the energy barriers for thermodynamically-favourable structural changes. According to the simple Jellium model [29], the double-anion Au_{16}^{2-} should be a closed-shell particle due to the 16 s-orbital electrons plus two excess electrons totalling a "magic number" of 18 electrons, and more accurate electronic structure (DFT) calculations have confirmed this view [25].

In this chapter, we report on density functional theory (DFT) calculations of soft-landed Au_{16} on a (0001) HOPG surface. There are many possible orientations of Au_{16} on the substrate (namely on vertices, edges and faces), of which we consider two: adsorption on the (111) and (0001) faces of the Au_{16} cluster (Fig. 4.1a,b, respectively). The substrate is modelled as a periodic slab of two graphene layers. The interaction between Au and the HOPG surface is weak physisorption [7]. However, it can be easily modified by defects, and we consider one, two and three neighbouring carbon vacancies (Fig. 4.1c). We report on the effect these changes have on Au–C interactions (chemisorption), as well as the significance on the geometric and electronic configurations of the Au_{16} cluster and HOPG surface. We also report tests for van der Waals-corrected exchange-correlation functionals in this context. The optimised Au_{16} geometries on HOPG and carbon nanotubes are further studied in order to probe the catalytic activity for CO oxidation. We also look at changes in the adsorption of Au_{16} on doped-HOPG surfaces.

4.2 Methodology

4.2.1 The Schrödinger Equation

The Schrödinger equation for a single particle of mass m moving in one dimension with energy E is

$$-\frac{\hbar^2}{2m}\frac{d^2\psi}{dx^2} + V(x,t) = E\psi \tag{4.1}$$

where ψ is the wavefunction of the particle, $\hbar$ is the reduced Planck's constant, and the potential energy, V, depends on the position x of the particle at time t [31].

For a multi-particle system of N electrons and M nuclei, we must in principle solve the expanded Schrödinger equation with the well known form:

$$\hat{H}\Psi(\boldsymbol{r}_1,\ldots,\boldsymbol{r}_N,\boldsymbol{R}_1,\ldots,\boldsymbol{R}_M) = E\Psi(\boldsymbol{r}_1,\ldots,\boldsymbol{r}_N,\boldsymbol{R}_1,\ldots,\boldsymbol{R}_M) \tag{4.2}$$

where $\hat{H}$ represents the Hamiltonian of the system, E is the energy of the eigenstate and Ψ is the wavefunction of said eigenstate [31]. The value of Ψ depends on the positions of both the electrons$(\boldsymbol{r}_1,\ldots,\boldsymbol{r}_N)$ and the nuclei $(\boldsymbol{R}_1,\ldots,\boldsymbol{R}_N)$ in the system.

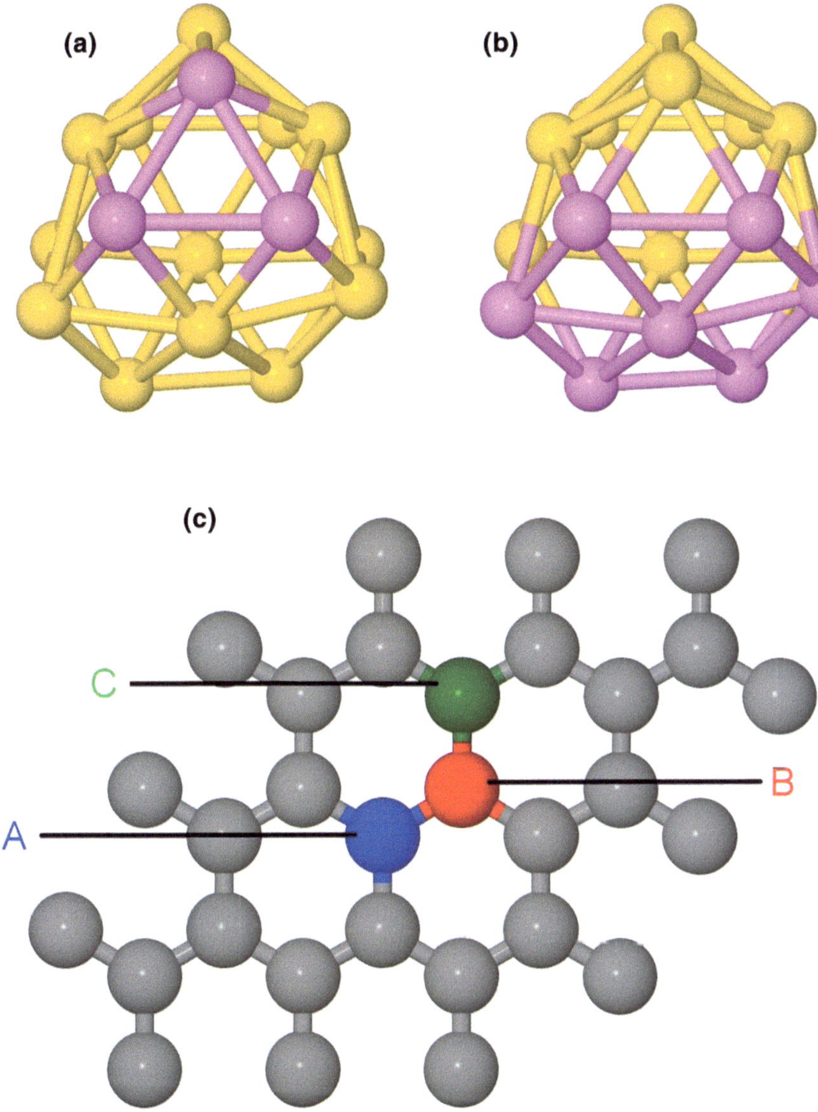

Fig. 4.1 *Top* to *bottom* **a** and **b** The T_d-symmetry Au_{16} cluster with the (111) and (0001) faces identified in *violet*, respectively. **c** A layer of HOPG, with one, two and three neighbouring surface defects identified as the removal of the atoms marked A, A and B, and A, B and C, respectively. Reprinted with permission from Ref. [30]. Copyright 2011 American Chemical Society

The Hamiltonian can be formed by the quantising of classical energies in Hamilton form. This molecular Hamiltonian is a sum of five parts:

$$\hat{H} = \hat{T}_e + \hat{T}_n + \hat{V}_{ee} + \hat{V}_{en} + \hat{V}_{nn} \tag{4.3}$$

where the kinetic energy of the electrons ($\hat{T}_e$) and nuclei ($\hat{T}_n$) are defined as:

$$\hat{T}_e = -\sum_i^N \frac{\hbar^2}{2m_e} \nabla^2 (\boldsymbol{r}_i) \tag{4.4}$$

$$\hat{T}_n = -\sum_i^M \frac{\hbar^2}{2M_i} \nabla^2 (\boldsymbol{R}_i) \tag{4.5}$$

where m_e is the mass of an electron and M_i is the mass of nucleus i. These kinetic terms are combined with terms for the potential energy arising from the repulsive Coulombic interactions between two electrons ($\hat{V}_{ee}$) and two nuclei ($\hat{V}_{nn}$), defined as:

$$\hat{V}_{ee} = \sum_i^N \sum_{j>i}^N \frac{e^2}{4\pi\varepsilon_0 |\boldsymbol{r}_i - \boldsymbol{r}_j|} \tag{4.6}$$

$$\hat{V}_{nn} = \sum_i^M \sum_{j>i}^M \frac{Z_i Z_j e^2}{4\pi\varepsilon_0 |\boldsymbol{R}_i - \boldsymbol{R}_j|} \tag{4.7}$$

and the Coulombic attraction between an electron and a nucleus ($\hat{V}_{en}$), defined as:

$$\hat{V}_{en} = -\sum_i^M \sum_j^N \frac{Z_i e^2}{4\pi\varepsilon_0 |\boldsymbol{R}_i - \boldsymbol{r}_j|} \tag{4.8}$$

where e is the elementary charge of an electron, Z_i is the atomic number of the nucleus i, and ε_0 is the permittivity of free space.

As the mass of the proton is three orders of magnitude greater than the mass of an electron, and the forces acting on the two particles are of similar magnitudes, it is reasonable to assume that the relaxation of electrons to their ground-state is instant with respect to the relaxation time of protons. The Born-Oppenheimer approximation [32] then allows the separation of the electronic (Ψ_e) and nuclear (Ψ_n) components of the wavefunction Ψ to give

$$\tilde{\Psi}(\boldsymbol{r}, \boldsymbol{R}) = \Psi_e(\boldsymbol{r}, \{\boldsymbol{R}\})\Psi_n(\{\boldsymbol{R}\}) \tag{4.9}$$

where $\{\boldsymbol{R}\}$ illustrates the use of the nuclear coordinates as a set of parameters. This allows a solution of the wavefunction whilst treating the nuclei as fixed point charges, after which the nuclei can be moved in the potential field generated by the electrons. With the removal of the nuclear components in the initial step, $\hat{H}$ becomes the electronic Hamiltonian $\hat{H}_{el} = \hat{T}_e + \hat{V}_{ee} + \hat{V}_{en}$. Thus the main task becomes solving the Schrödinger equation for electrons:

$$\hat{H}_{el}\Psi(\boldsymbol{r},\{\boldsymbol{R}\}) = E_{el}\Psi(\boldsymbol{r},\{\boldsymbol{R}\}) \tag{4.10}$$

and then later on the nuclear energies can be added to the electron energies, $\hat{H} = \hat{H}_{el} + \hat{T}_n + \hat{V}_{nn}$, giving a good approximation to the total energy of the system.

4.2.2 Density Functional Theory

Density Functional Theory, or DFT, looks to reformulate the Schrödinger equation in a philosophically and computationally different manner, using the theorems of Hohenberg and Kohn as its foundation [33]. The Coulombic attraction of electrons to a nuclear force, $\hat{V}_{en}$, can be regarded as the response of electrons to a constant external field, V_{ext}, and thus the electronic Hamiltonian can be rewritten:

$$\hat{H} = \hat{F} + \hat{V}_{ext} \tag{4.11}$$

where $\hat{F} = \hat{T}_e + \hat{U}_{ee}$ and

$$\hat{V}_{ext} = -\sum_{i}^{N} V_{ext}(\boldsymbol{r}_i) \tag{4.12}$$

$\hat{F}$ is the same for all N-electron systems, therefore $\hat{H}$, and consequently Ψ, depend only on N and $V_{ext}(\boldsymbol{r})$. V_{ext} proves to be a functional of the electron density, $n(\boldsymbol{r})$: this is the first of the Hohenberg-Kohn theorems. Furthermore, knowledge of the ground-state electron density $n_0(\boldsymbol{r})$ and N, where:

$$N = \int n_0(\boldsymbol{r})d\boldsymbol{r} \tag{4.13}$$

can be used to calculate any ground-state property, including $\Psi[n_0]$. The second Hohenberg-Kohn theorem states that there is a functional of electron density, $E[n]$, which gives the energy of a system and the ground-state electron density minimises this to give the ground-state energy, i.e. $E[n_0] = E_0$. This ground-state can be found via variational principle as a functional of the ground-state electron density:

$$E_0 \leq E[n] = \langle\Psi|\hat{H}|\Psi\rangle = \langle\Psi[n]|\hat{F} + \hat{V}_{ext}|\Psi[n]\rangle = F[n] + \int n(\boldsymbol{r})V_{ext}(\boldsymbol{r})d^3\boldsymbol{r} \tag{4.14}$$

where $F[n] = \langle\Psi|\hat{F}|\Psi\rangle$ and $\langle\Psi|\hat{V}_{ext}|\Psi\rangle = \int n(\boldsymbol{r})V_{ext}(\boldsymbol{r})d^3r$.

Whilst this proves the existence of the functional $F[n]$, it does not give the form. From the Hohenberg-Kohn theorems, Kohn and Sham derived a set of equations for a non-interacting system of N-electrons from which, in principle, n could be calculated equivalent to the same system of interacting particles [34]. In Kohn-Sham (KS) theory, this is formulated as a simple expression for the ground-state energy [35]:

$$E[n] = T_s[n] + V_{en}[n] + J[n] + E_{xc}[n] \tag{4.15}$$

where the form of some of the functionals is explicitly known: V_{en} is the nucleus-electron potential energy, expressed in terms of an external potential due to the nuclei, V_{ext}, as discussed above. $T_s[n]$ is the KS kinetic energy of the non-interacting electrons, given as:

$$T_s[n] = \langle\psi|T_e|\psi\rangle = -\sum_i^N \int \psi_i^*(\boldsymbol{r}) \left(\frac{\hbar}{2m_e}\nabla^2\right) \psi_i(\boldsymbol{r}) d\boldsymbol{r} \tag{4.16}$$

expressed in terms of $\{\psi_i\}$, the set of one electron KS orbitals. The other known energy component is the classical electron–electron repulsion energy, or Coulomb energy:

$$J[n] = \frac{e^2}{2} \int\int \frac{n(\boldsymbol{r})n(\boldsymbol{r}')}{|\boldsymbol{r} - \boldsymbol{r}'|} d\boldsymbol{r} d\boldsymbol{r}' \tag{4.17}$$

Much is known about the key remaining term, the exchange-correlation functional $E_{xc}[n]$, though no explicit form can be defined: The role of the exchange-correlation functional is to correct for the approximations made in the non-interacting particle assumptions of KS theory.

The electron density of the system is constructed as a sum over particle indices:

$$n(\boldsymbol{r}) = \sum_i^N |\psi_i(\boldsymbol{r})|^2 \tag{4.18}$$

The one electron KS orbitals, $\{\psi_i\}$, in Eqs. (4.16) and (4.18) are constructed from the KS equation for the one electron orbitals:

$$\left(-\frac{\hbar}{2m_e}\nabla^2 + V_{KS}(\boldsymbol{r})\right) \psi_i(\boldsymbol{r}) = \varepsilon_i \psi_i(\boldsymbol{r}) \tag{4.19}$$

where ε_i corresponds to the energy of the KS orbital ψ_i, $V_{KS}(\boldsymbol{r})$ is the Kohn-Sham potential:

$$V_{KS} = e^2 \int \frac{n(\boldsymbol{r}')}{|\boldsymbol{r} - \boldsymbol{r}'|} d\boldsymbol{r}' + V_{ext}(\boldsymbol{r}) + V_{xc}(\boldsymbol{r}) \tag{4.20}$$

and V_{xc}, the exchange-correlation potential, is the functional derivative of the exchange-correlation energy:

$$V_{xc}(\boldsymbol{r}) = \frac{\delta E_{xc}[n]}{\delta n(\boldsymbol{r})} \tag{4.21}$$

The KS equations are solved in a self-consistent fashion. n is initially guessed, and then using an approximate form for the functional dependence of E_{xc}, $V_{KS}(\boldsymbol{r})$ is computed. The KS equations are then solved to give an initial estimate of $\{\psi_i\}$ (Eq. 4.19). This set of orbitals is used to calculate an improved density using Eq. (4.18), and the process repeats until convergence. The final energy is calculated from Eq. (4.15).

It must be emphasised that DFT is an exact theory for finding the ground-state properties of an N-electron system; however practical approximations are necessary to take into consideration the exchange and correlation effects represented by E_{xc}. These approximations define the accuracy with which structural and energetic properties of a ground-state system are calculated. Unless stated otherwise, the generalised gradient approximation (GGA) is used for E_{xc} in this work:

$$E_{xc} = E_{xc}^{GGA}[n(\boldsymbol{r}), \nabla n(\boldsymbol{r})] \tag{4.22}$$

which is dependent on both the electron density, $n(\boldsymbol{r})$, and also the gradient of the density.

4.2.3 The Projector Augmented Wave Method

Chemical properties are characteristically defined by the valence electron configuration. For elements in the third period of the periodic table, and higher, there are a larger number of core electrons which are unimportant in a chemical sense; however without them the valence electrons cannot be truly represented and thus a large number of basis functions are required to expand the necessary orbitals [36]. Furthermore, relativistic effects also have an effect on elements in the lower half of the periodic table. These computational complications can be treated by the use of pseudopotentials to represent the core and valence electrons separately, with only the latter treated explicitly; the smoother wavefunctions significantly reduce computing time [31]. Pseudopotentials replace the strong Coulomb potential between the nucleus and valence electrons inside a pre-defined core with a weaker, effective potential, taking into consideration the screening effect core electrons have on the nucleus, as shown in Fig. 4.2. Beyond the pre-defined core-cutoff, pseudopotentials and the total potential are equivalent [31].

The projector-augmented wave (PAW) method uses a similar idea as that of pseudopotentials to reduce computational requirements [37, 38], whilst conserving the true wavefunction, ψ.

In the PAW approach a KS all-electron wavefunction, $|\psi\rangle$, which oscillate rapidly near an atoms nuclear core, are transformed into a smoother pseudo-wavefunction, $|\tilde{\psi}\rangle$, by a linear transformation operator $\hat{T}$:

$$|\psi\rangle = \hat{T}|\tilde{\psi}\rangle \tag{4.23}$$

Around each nucleus a, with position $\boldsymbol{R}^a$, a cutoff r_c^a is defined. r_c^a has a similar meaning to the core-cutoff used for pseudopotentials: beyond this cutoff the valence

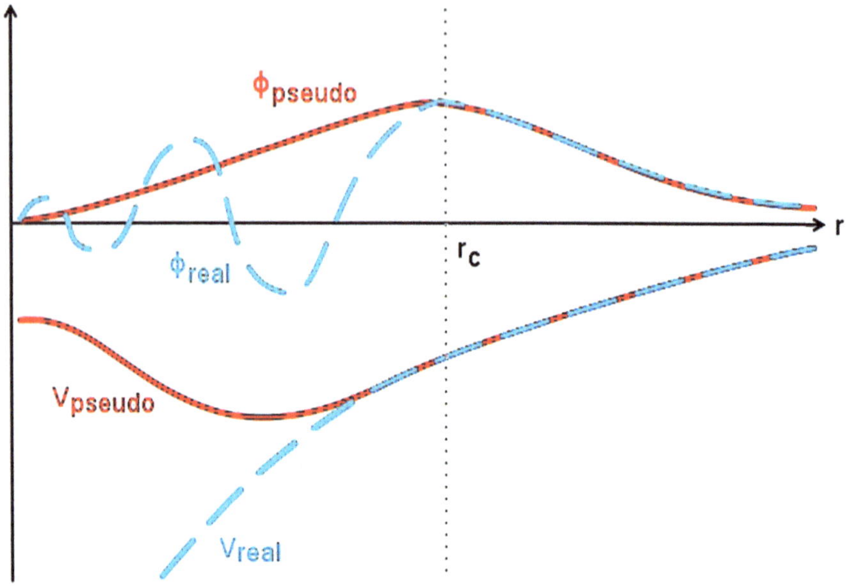

Fig. 4.2 Comparison of a partial wave (ϕ) in the Coulomb potential of a nucleus (*blue*) to one in a pseudopotential (*red*). The real and pseudo partial waves, and potential energies V, match beyond the cutoff radius r_c

wavefunction is smooth, thus continuity is expected where $|\boldsymbol{r} - \boldsymbol{R}^a| \geq r_c^a$ such that $|\psi\rangle = |\tilde{\psi}\rangle$. Localised within the augmentation spheres ($|\boldsymbol{r} - \boldsymbol{R}^a| < r_c^a$) it is necessary to expand $|\psi\rangle$ in to a complete basis of i partial waves, $\phi(\boldsymbol{r})$. Smooth projector functions, $\tilde{p}_i^a$, act on the $\phi(\boldsymbol{r})$, creating smooth partial waves, $\tilde{\phi}(\boldsymbol{r})$:

$$\hat{T} = 1 + \sum_a \sum_i (|\phi_i^a\rangle - |\tilde{\phi}_i^a\rangle)\langle\tilde{p}_i^a| \qquad (4.24)$$

In summary, the all-electron and pseudo-wavefunctions differ within the core augmentation sphere, however they match exactly in the valence region. The partial waves and projection operators do not depend on the system being studied, and thus only need be calculated once for each element. The total PAW energy is calculable as:

$$E = \tilde{E} + \sum_a (E^a - \tilde{E}^a) \qquad (4.25)$$

where $\tilde{E}$ is the "soft" energy contribution evaluated outside all augmentation spheres, which is calculated on a 3D grid, whilst E^a and $\tilde{E}^a$ are calculated within the core region on a radial grid. This scheme allows the combination of energies from two traditionally incompatible grids [38].

The all-electron density is obtained as:

$$n(\boldsymbol{r}) = \tilde{n}(\boldsymbol{r}) + \sum_a (n^a(|\boldsymbol{r} - \boldsymbol{R}^a|) - \tilde{n}^a(|\boldsymbol{r} - \boldsymbol{R}^a|)) \tag{4.26}$$

where $\tilde{n}(\boldsymbol{r})$ is the pseudoelectron density, $n^a(\boldsymbol{r})$ is the atom-centered all-electron density and $\tilde{n}^a(\boldsymbol{r})$ is the atom-centered pseudodensity.

4.3 Simulation Methods

The calculations presented have been performed with the real-space finite-difference DFT code GPAW [38, 39]. GPAW uses the all-electron density, where the electrons are described using the PAW method in the frozen core approximation [37]. The valence electrons are taken as the outermost 11 and 4 electrons for Au ($5d^{10}6s^1$) and C ($2s^22p^2$), respectively. The PAW setup for the Au valence configuration includes scalar-relativistic effects. The GGA of Perdew, Burke and Ernzerhof (PBE) [40] is used for the exchange-correlation energy functional unless otherwise indicated. The KS wave functions are expanded directly on a real space grid, with grid spacing of 0.2 Å. The Hamiltonian is rediagonalised after each geometry step using the non-default Davidson eigensolver [41], and spin-paired calculations are performed unless explicitly mentioned otherwise.

Our substrate of Bernal graphite comprises two graphene layers (AB stacking) in a hexagonal supercell of $14.76 \times 14.76 \times 20$ Å^3 (144 C atoms), with periodic boundary conditions employed in the x- and y-directions. The choice of unit cell keeps the Au_{16} adsorbates approximately 9 Å apart laterally. Previous calculations show little difference between the results for atomic Au adsorption on to two and three graphene layers [7], making it acceptable as an approximation to multi-layered graphite. The bond lengths and interlayer spacing of C are initially fixed to experimental values of 1.42 and 3.35 Å, respectively [42], as the PBE functional does not describe weak dispersion forces between graphene layers well [43]; some calculations are repeated for comparison using the van der Waals-corrected vdW-DF [44] and vdW-DF2 [45] exchange-correlation functionals, which represent these interactions better.

The effect of surface relaxation has been incorporated throughout the calculations by releasing constraints on C atoms within a specified starting distance of the Au cluster, namely 3.5 and 7.0 Å (totalling 14–23 and 94–105 atoms, respectively, and depending on cluster orientation). The fixed graphene layer spacing causes an effective repulsion, evident as a small bending of the released substrate area. This feature occurs also for vdW-corrected calculations, which indicates that it is caused primarily by the relaxation of the intralayer C–C bonds, not by the interlayer repulsion of graphene sheets. The ionic positions are optimised using a quasi-Newtonian (BFGS) method [46] until the maximal force for unconstrained atoms is below 0.02 eV/Å, and the total energy change between three steps is less than 1×10^{-4} eV. The KS orbitals are re-optimised after each step, with the change in density less than 1×10^{-5}

electrons per valence electron and the integrated value of the square of the residuals of the KS equations less than 1×10^{-11} per state. Charge transfer is calculated using the Bader charge analysis as implemented by Tang et al. [47], and electron localisation function (ELF) [48] analysis is calculated directly from GPAW.

All projected local densities of states (PLDOS) are computed by directly projecting wavefunctions on to atomic orbitals centred at minimised coordinates. For a smoother DOS profile, it is necessary to repeat DFT calculations with a $5 \times 5 \times 1$ Monkhorst-Pack **k**-point mesh, corresponding to 13 explicit **k**-points in the Brillouin zone, allowing denser sampling of the energy states within the HOPG structure [49]. As a by-product, we are able to compare E_{tot} for systems after the 13 **k**-point calculations to those of our results from the structural Γ-point calculations, thus allowing us to check the validity of the Γ-point approximation. The difference between E_{tot} for the levels of computation (ΔE_{tot}) is given in Appendix C.1. We find that $2.79\,\text{eV} > \Delta E_{tot} > 1.96\,\text{eV}$, a difference in E_{tot} of 0.21–0.15 %; this value is relatively small, and consistent in size. The norm of the forces on carbon atoms is within 0.001 eV/Å for the two approaches justifying our choice of the Γ-point for structural calculations.

4.4 HOPG Surface

Au_{16} coordinates were obtained from previous studies [25] and relaxed. The cluster binding energy (E_b) is calculated to be 2.30 eV per atom (bulk cohesive energy is 3.81 eV per atom [50]) and a HOMO-LUMO gap of 0.17 eV. The energy gap between the LUMO and the next unoccupied orbital (LUMO + 1) is 1.28 eV, a strong reason for the anionic favourability of this structure (our system comprises doubly-degenerate orbitals). The number of interatomic interactions below 3.5 Å in length ($N_{\text{Au–Au}}$) is 42, for which the average (mean) bond length ($\bar{r}_{\text{Au–Au}}$) is 2.80 Å, with a range of 2.69–2.99 Å depending on the coordination.

4.4.1 Adsorption Interactions

Results of Au_{16} adsorption on a graphite substrate with and without surface vacancies, with C atoms unconstrained within 3.5 Å of the Au_{16} cluster, are shown in Table 4.1. Adsorption energy (E_a) is a measure of the interaction between adsorbate and substrate and is calculated as:

$$E_a = E^f_{ads} + E^f_{sub} - E_{tot} \tag{4.27}$$

where E^f_{ads} and E^f_{sub} are the energies of the structurally minimised isolated (i.e. "free") adsorbate (cluster) and substrate (surface), respectively, and E_{tot} is the total system energy when combined. This gives us a measure of surface–cluster interactions, taking into consideration structural rearrangements.

Results for adsorption on the (111) cluster face show increasing E_a for a greater number of surface vacancies. The strongest binding (5.06 eV) is calculated for a HOPG surface with three neighbouring vacancies (GR2-3h). Strong binding (3.79 and 3.07 eV) is also calculated for a graphite surface with two neighbouring vacancies (GR2-2h) and a similar system with only a single atom defect (GR2-h), respectively. A defect-free surface (GR2) forms only weak interactions (0.80 eV) with Au_{16}. Previous DFT studies report $E_a = 0.25–0.80$ eV for Au atoms and small clusters adsorbed on HOPG [7, 51]. Calculations with the vdW-DF and vdW-DF2 functionals (Appendix C.2) increase E_a significantly (up to 80 %) on GR2 (1.35 and 1.47 eV, respectively) due to the physicality of the long-range interaction. However, for adsorption on to GR2-2h, E_a (vdW-DF) and E_a (vdW-DF2) are 3.10 and 2.91 eV, respectively, which is lower than for the PBE functional: this is coupled to relaxation of the adsorption site, as discussed below. E_a on GR2-2h is large for PBE, vdW-DF and vdW-DF2 compared to GR2 as chemical interactions dominate in the presence of a surface defect.

The observation that Au_{16} binds weakly to HOPG (GR2) is reinforced by the relatively large Au-X interaction distance (3.33 Å), compared to 1.95–2.09 Å in the presence of a surface defect (GR2-h, GR-2h and GR2-3h). The vdW-DF and vdW-DF2 functionals give slightly decreased Au-X distances on GR2 (3.32 and 3.25 Å, respectively), but quantitatively similar distances when a defect is present (±0.01 Å). The situation is different for benchmark calculations of the optimal graphene layer separation, where the vdW-DF and vdW-DF2 functionals result in improved values of 3.50 and 3.40 Å (PBE result 3.75 Å) for two layers.

Attention to the cumulative charge transfer (Δq) of the Au_{16} cluster, for GR2, shows a total of 0.75 e, emphasising the presence of strong polarisation. We should remind ourselves here that the T_d-symmetry Au_{16} structure is the anionic global minimum, and so one would expect the neutral structure to be electronegative ($\Delta q > 0$) in these circumstances as it draws electron density from the HOPG surface. Δq is significantly large (0.51 e) for GR2-h also, whilst for GR2-2h and GR2-3h the charge transfer to Au becomes slightly negative (−0.05 to −0.06 e), i.e. towards the C atoms.

Analysis of the change in atomic charge levels helps identify the reason for this change in Δq. The change in localised charge (Δq_{atom}) of the Au atom closest to the HOPG-plane is found to have a strong influence on the Δq value for each adsorbed cluster, as can be seen clearly in Fig. 4.3: Δq_{atom} values of 0.06, −0.30, −0.43 and −0.40 e for GR2, GR2-h, GR2-2h and GR2-3h, respectively. We note that the Au atom second closest to the GR2-3h HOPG surface is also positively charged ($\Delta q_{atom} = -0.21$ e), and electron density has been transferred onto neighbouring C atoms for both these Au atoms.

Attention to the vertical displacement of the lowest Au from the top graphite plane [$d_{\perp}$(Au)] shows that the observed Δq neutrality for GR2-2h and GR2-3h is coupled with the Au atom being close to the plane of the graphite surface (0.51 and 0.78 Å, respectively). For GR2, the opposite is observed as the strong electronegativity

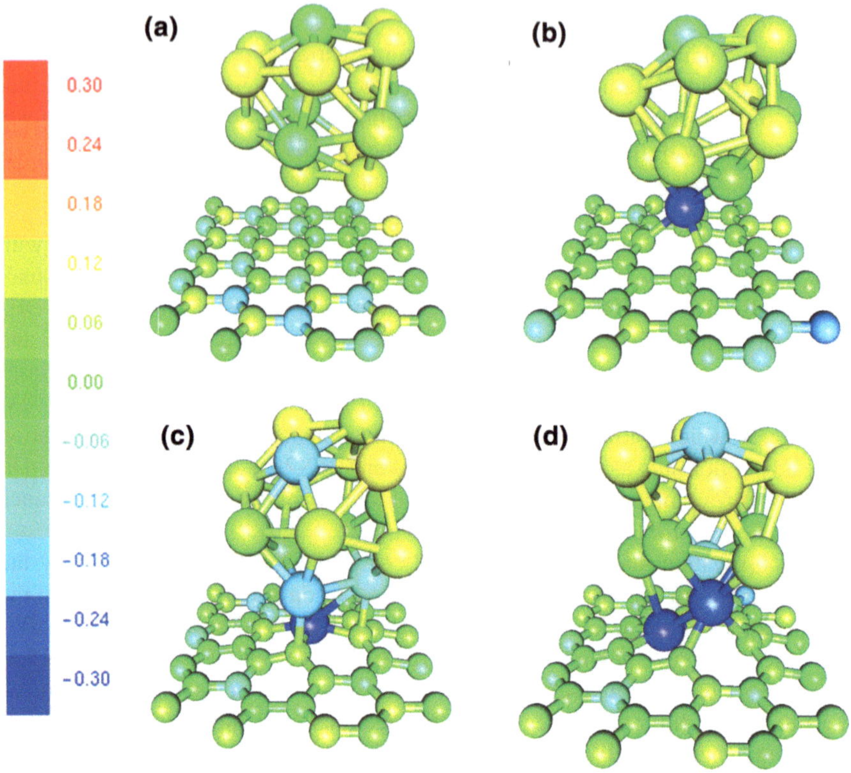

Fig. 4.3 Δq_{atom} illustrated by colour superimposed on the coordinates of the DFT-calculated minima for Au_{16} adsorbed on the (111) face for **a** GR2, **b** GR2-h, **c** GR2-2h and **d** GR2-3h. A key is provided on the *left*, where *red* illustrates electron charge transfer to an atom (i.e. negatively charged), *green* is no change and *blue* is transfer from the atom (i.e. positively charged). C atoms can be identified as those lying in-plane initially, with an illustrative radius half that of the Au atoms. Reprinted with permission from Ref. [30]. Copyright 2011 American Chemical Society

couples with large $d_{\perp}$(Au) (3.38 Å). GR2-h falls between these two values (1.78 Å), as the Au atom is unable to penetrate the HOPG surface; this is consistent with previous observations for Au adatoms on graphene [11]. The displacement of the underlying substrate atom [$d_{\perp}$(X)] to the Au cluster also shows a similar trend. For GR2, $d_{\perp}$(X) is small (0.07 Å), but when a vacancy appears (GR2-h) $d_{\perp}$(X) increases 6-fold (0.42 Å). GR2-2h and GR2-3h have $d_{\perp}$(X) of 0.58 Å and 0.61 Å, respectively. Calculations for the isolated GR2-h, GR2-2h and GR2-3h give $d_{\perp}$(X) of 0.06, 0.06 and 0.05 Å, respectively, showing that this displacement, and associated bending of the HOPG plane, is induced by the cluster as a chemical, and not physical, process. Further analysis with the vdW-DF and vdW-DF2 functionals confirms that this is the case (see Appendix C.3).

Results for adsorption on the Au_{16} (0001) face are also listed in Table 4.1. As the number of defects increase (GR2 → GR2-h → GR2-2h → GR2-3h), E_a and $d_{\perp}$(X)

Table 4.1 Au_{16} cluster on HOPG with and without multiple surface defects: perfect graphite (GR2), one surface vacancy (GR2-h), two surface vacancies (GR-2h) and three surface vacancies (GR2-3h)

		GR2	GR2-h	GR2-2h	GR2-3h
Au (111) face:					
E_a	(eV)	0.80	3.07	3.79	5.06
Au-X	(Å)	3.33	2.09	1.99	1.95
$d_\perp$(Au)	(Å)	3.38	1.78	0.78	0.51
$d_\perp$(X)	(Å)	0.07	0.42	0.58	0.61
Δq	(e)	0.75	0.51	−0.06	−0.05
Au (0001) face:					
E_a		0.78	3.07	3.16	4.63
Au-X		3.60	2.09	1.98	1.96
$d_\perp$(Au)		3.58	1.85	0.50	0.59
$d_\perp$(X)		0.07	0.46	0.59	0.65
Δq		0.76	0.44	0.05	0.01

In all cases X is considered to be the nearest neighbouring C atom to the Au_{16} cluster, and all C atoms are unconstrained within 3.5 Å. Adsorption energy, (E_a), vertical displacements from the upper graphene plane [$d_\perp$(Au) and $d_\perp$(X)], the nearest neighbour (Au-X) distances, and the charge transfer (Δq) are given. $\Delta q > 0$ implies charge transfer to the Au cluster

increase (0.78–4.63 eV and 0.07–0.65 Å, respectively), whilst Au-X, $d_\perp$(Au) and Δq decrease (3.6–1.96 Å, 3.58–0.59 Å and 0.76–0.01 e, respectively), as illustrated in Fig. 4.4. This is consistent with the previously observed trend for adsorption on the Au_{16} (111) face, that as surface vacancies increase the T_d-symmetry of the Au_{16} structure tends away from its electronegative starting coordinates, with the graphene plane and cluster attracted together. For GR2-2h and GR2-3h $d_\perp(X) > d_\perp(Au)$, showing that the lowest Au is already in the HOPG layer.

GR2-h, GR2-2h and GR2-3h have three, four and five "dangling" bonds, respectively, and this influences the adsorption bonding and cluster positioning on the HOPG surface. In all cases in our calculations all the available "dangling" bonds from the defected graphite plane participate in absorbing the Au_{16} cluster. For GR2-h, an Au atom is positioned in a "tripod" position over the single defect [the positioning is akin to adsorption on the (111) face], whilst for GR2-2h there is space for an Au atom to be absorbed in plane (Fig. 4.5). Such is the preference for in-plane adsorption of an Au atom that the Au_{16} cluster is stretched apart, most noticeably when absorbed on the (0001) face (Fig. 4.5a). This displacement of an Au atom from the nanoparticle into the HOPG plane is structurally unfavourable with respect to the Au_{16} cluster due to the required weakening of Au–Au bonds, and explains the limited change seen in E_a between GR2-h and GR2-2h for (0001) adsorption despite the Au_{16} cluster's greater proximity to the HOPG surface in the latter. For GR2-3h, we see an Au atom adsorbed in plane, similar to GR2-2h, and then a neighbouring Au atom from the cluster taking up the "tripod" position over the remaining vacancy, using the in-plane Au as one of the three base positions (Fig. 4.5b). This results in a significant increase in E_a, compared to GR2-2h, due to increased structural stability.

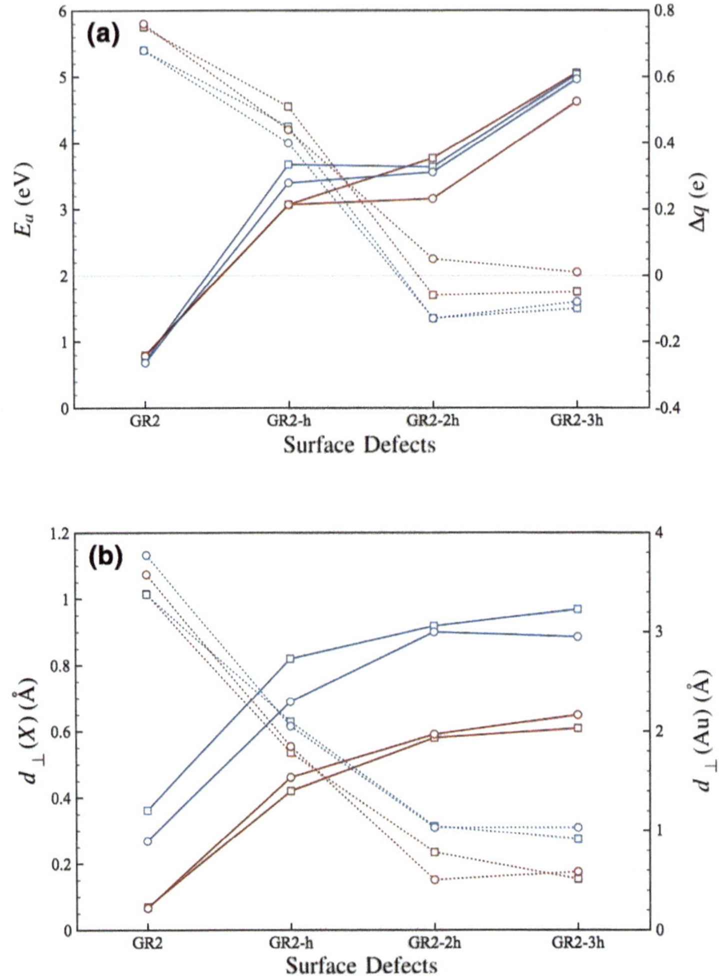

Fig. 4.4 *Top* to *bottom* Surface Defects (**a**) E_a (*solid line*, left y-axis) plotted alongside Δq (*dashed line*, *right* y-axis) against number of surface defects and (**b**) $d_\perp$(X) (*solid line*, *left* y-axis) plotted with $d_\perp$(Au) (*dashed line*, *right* y-axis) against number of surface defects. *Squares* and *circles* represent adsorption on Au_{16} (111) and (0001), respectively. *Red* and *blue colouring* represent a 3.5 and 7.0 Å surrounding distance of unconstrained C atoms from the Au_{16} cluster, respectively. Reprinted with permission from Ref. [30]. Copyright 2011 American Chemical Society

Comparison between the results for Au_{16} adsorption on the (111) and (0001) faces raises some interesting observations: When an Au atom is in the HOPG plane, E_a is greater for adsorption on the Au_{16} (111) face for GR2-2h (3.79–3.16 eV) and GR2-3h (5.06–4.63 eV). This is due to the greater distortion of the Au_{16} cluster when absorbed on to its (0001) face, as previously mentioned. Pivotal to this observed distortion is the attraction of an Au atom to the vacancy of size two (GR2-2h) or

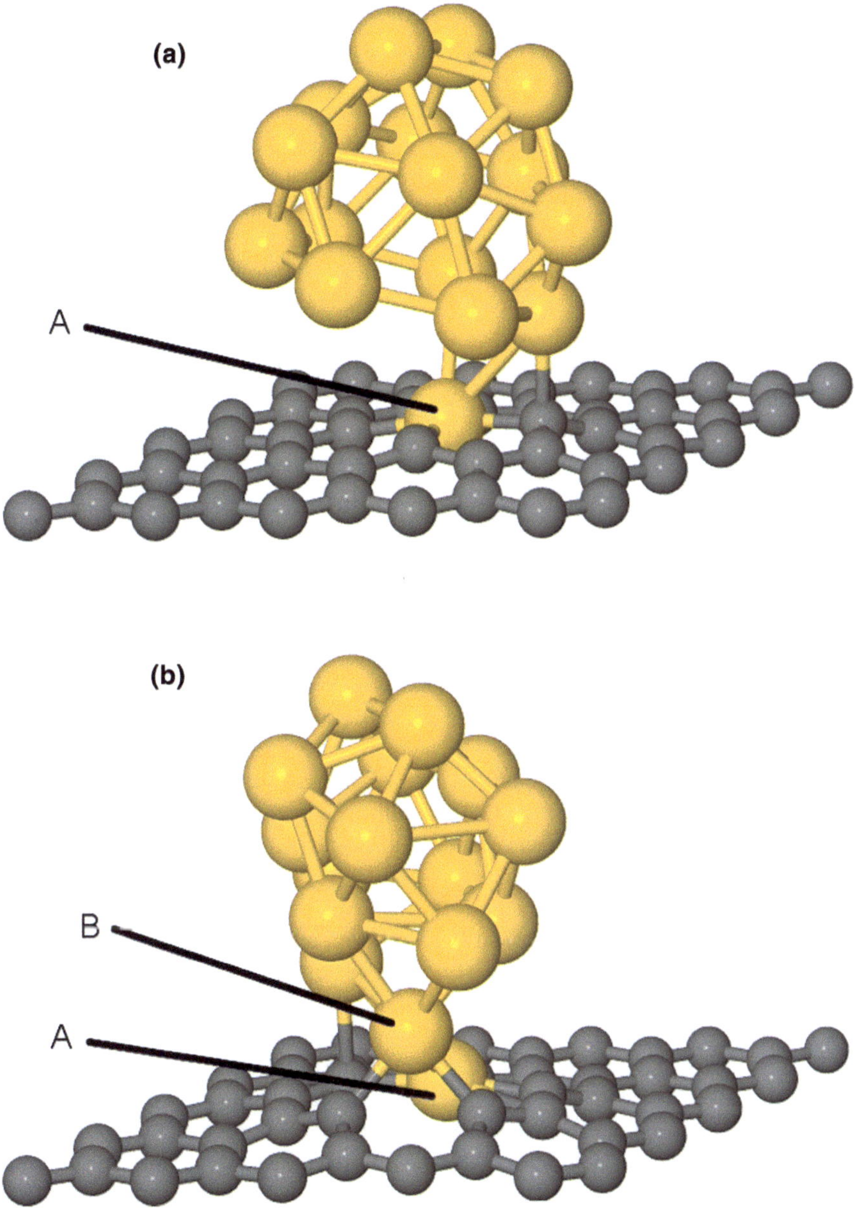

Fig. 4.5 *Top* to *Bottom* Au_{16} deposited on (0001) face on (**a**) GR2-2h and (**b**) GR2-3h. For the purpose of illustration extraneous C atoms have been removed. Labels A and B identify atoms adsorbed in-plane and in a "tripod" position, respectively. Reprinted with permission from Ref. [30]. Copyright 2011 American Chemical Society

greater in the HOPG plane (Fig. 4.5), coordinating in-plane to the four "dangling" bonds, and thus completing the plane and minimising further attraction between the out-of-plane Au atoms and the HOPG layer.

Despite increased E_a, Au-X is the same for GR2-h, GR2-2h and GR2-3h ($\pm$0.01 Å) in both cases. The vertical separation is larger for adsorption on the Au_{16} (0001) face on to GR2 (3.33–3.60 Å), but these are both large, and thus weak, interaction distances of physical character only. $d_{\perp}$(Au) is greater for adsorption on the Au_{16} (0001) face for GR2, GR2-h and GR2-3h. For GR2-2h the Au_{16} (0001) adsorption forms a particularly stable HOPG orientation, with low $d_{\perp}$(Au) due to strong in-plane Au atomic adsorption. Previous work cites two defects as the optimum size vacancy for a solitary Au adatom on graphene [11]. The observed trend for $d_{\perp}$(X) is consistent ($\pm$0.04 Å) for GR2-2h and GR2-3h, where electron–neutral interactions are formed with the HOPG substrate ($\Delta q \approx 0$).

The in-plane adsorption of an Au atom from the Au_{16} cluster for GR2-2h (Fig. 4.5) is of further interest. The electron localisation function (ELF) reflects the probability of finding two electrons at the same location, thus giving an illustrative measure of the type of bonding present. ELF calculations for adsorption of the Au_{16} (0001) face on to GR2-2h are given in Fig. 4.6. Scaling runs from 0 (no localisation and/or no electron density, blue) to 1 (complete localisation, red). A reference value of 0.5 (green) corresponds to a homogeneous electron gas (metallic bonding), whilst covalent bonds appear red due to the electron pairing.

ELF shows the covalent nature of the C–C bond (dark red), and for Au–Au we see delocalised electrons (green) around the nucleus with a larger radius than for C due to the greater number of electrons. The metallic Au–Au bonds are not visible due to the irregularity of the bond alignments to the principle planes (xy-, xz- and yz-), but these are present in the system. A metallic-type bond is also formed between Au–C with delocalisation (green colour) visible between the in-plane Au atom and neighbouring C atoms (Fig. 4.6a); however there is no chemical bonding between any out-of-plane Au and HOPG C atoms (Fig. 4.6b). The bent nature of the HOPG top-layer can be identified as C atoms that lie above and below the cut-plane, depending on their proximity to the Au_{16} cluster (Fig. 4.6a). There is no significant deformation of the ELF for the top layer of C atoms in GR2, indicating that chemical bonding has not changed considerably within the graphene plane.

Results of Au_{16} adsorption on an HOPG substrate with and without surface vacancies, with C atoms unconstrained within 7.0 Å of the Au_{16} cluster, are given in Appendix C.5. General trends are as previously observed for adsorption on to a surface with an unconstrained distance of 3.5 Å (Table 4.1). The increased freedom of the C atoms results in minimal changes to the range of E_a, and thus the hierarchy of the different cluster-substrate interactions. The only relational difference is the decrease in E_a for Au_{16} (111) face adsorption on GR2-2h when compared to GR2-h, which is noted as being due to distortion of the starting T_d-symmetry geometry. $d_{\perp}$(X) increases throughout due to the greater area of C free movement, directly leading to decreased $d_{\perp}$(Au) values as the vertical displacement of the top HOPG layers increases. Though an Au atom still sits in plane for calculations on GR2-2h and GR2-3h, there is an increased bending in the topmost HOPG layer; and thus greater

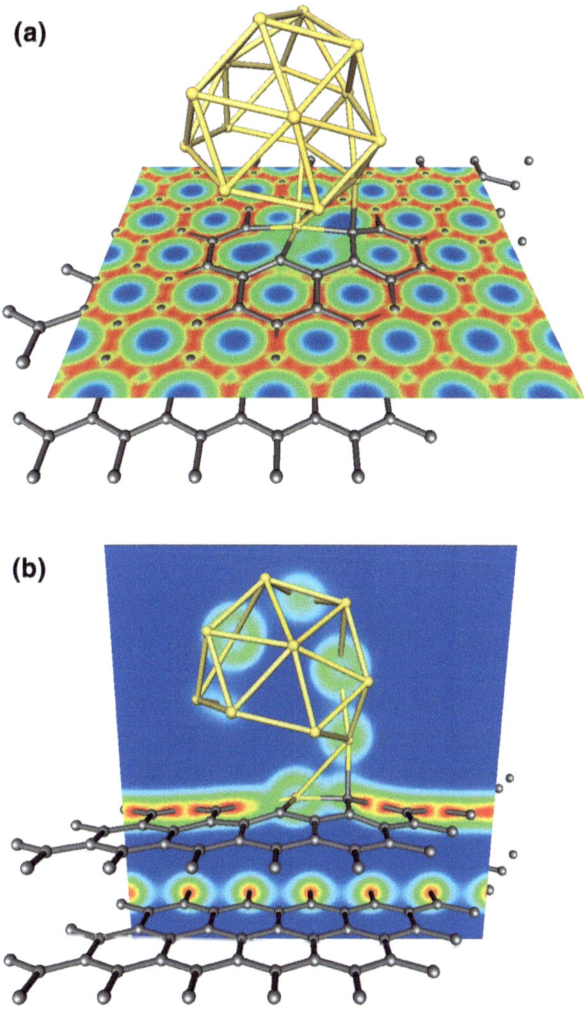

Fig. 4.6 *Top* to *bottom* Cutplane presentation of the ELF for Au_{16} deposited on (0001) face on GR2-2h, in both the (**a**) xy-plane and (**b**) xz-plane. The *red colour* corresponds to full localisation (1.0, covalent bonds), *green* is analogous to homogeneous electron gas (0.5, metallic bonding), and *blue* equals low localisation (0.0). Reprinted with permission from Ref. [30]. Copyright 2011 American Chemical Society

displacement. The increased planar bending also effects Δq, with values more negative across the board (GR2, GR2-h, GR2-2h and GR2-3h). Calculated interlayer distances and $d_\perp$ (X) for the isolated HOPG systems with both 3.5 and 7.0 Å of unconstrained C atoms are given in Appendix C.5, reaffirming the observation that the surface bending is predominantly a cluster-induced effect.

We note that previous observations have shown a surface of GR2-h type to have a local magnetic moment, i.e. triplet state around the defect [52]. Spin-unpaired calculations, with an initial magnetic moment, were conducted on GR2-h with an unconstrained radius of 7.0 Å, and this gave rise to a lower E_{sub} by 0.43 eV. Expansion of two distant C–C interactions and contraction of one C–C interaction (2.12 Å) across the vacancy were noted, in line with previous results (geometry shown in Appendix C.4). Au_{16} calculations with spin-unpaired configuration give negligible changes to E_{ads}, and therefore, one can decrease the documented E_a for GR2-h by the above value to take this spin-unpaired effect of the substrate into consideration.

4.4.2 Structural Observations

Structural observations of the adsorbed Au_{16} cluster with HOPG atoms constrained beyond 3.5 Å from the cluster are shown in Table 4.2, with images also given in Fig. 4.7. To aid structural analysis, a measure of the structural stability of the Au_{16} cluster in its adsorbed geometry is defined:

$$\Delta E = E^{f}_{ads} - E^{i}_{ads} \tag{4.28}$$

where E^{f}_{ads} is the energy of the relaxed gas-phase geometry, as defined in Eq. (4.27), and E^{i}_{ads} is the energy of the adsorbed Au_{16} cluster in an isolated (gas-phase) environment, without structural relaxation. We have an idea of the favourability

Table 4.2 Structural analysis of Au_{16} cluster on HOPG graphite with and without multiple surface defects as defined for 4.1

		GR2	GR2-h	GR2-2h	GR2-3h
Au (111) face:					
$N_{\text{Au-Au}}$		42	42	42	43
$\bar{r}_{\text{Au-Au}}$	(Å)	2.80	2.80	2.85	2.84
Symmetry		T_d	C_1	C_1	C_1
ΔE	(eV)	0.01	0.09	−0.42	−0.20
Au (0001) face:					
$N_{\text{Au-Au}}$		42	42	38	41
$\bar{r}_{\text{Au-Au}}$	(Å)	2.81	2.82	2.81	2.83
Symmetry		T_d	C_1	C_s	C_2
ΔE	(eV)	0.00	−0.06	−1.43	−0.77

C atoms are unconstrained within 3.5 Å of the Au_{16} cluster. The number of Au–Au bonds below 3.5 Å in length ($N_{\text{Au-Au}}$), average (mean) Au–Au bond length ($\bar{r}_{\text{Au-Au}}$) for all bonds in $N_{\text{Au-Au}}$, cluster symmetry and the difference in energy between the isolated Au_{16} structure (T_d symmetry) and the energy of the isolated Au_{16} cluster as adsorbed (ΔE) are given, where $\Delta E > 0$ implies the adsorbed structure is more stable

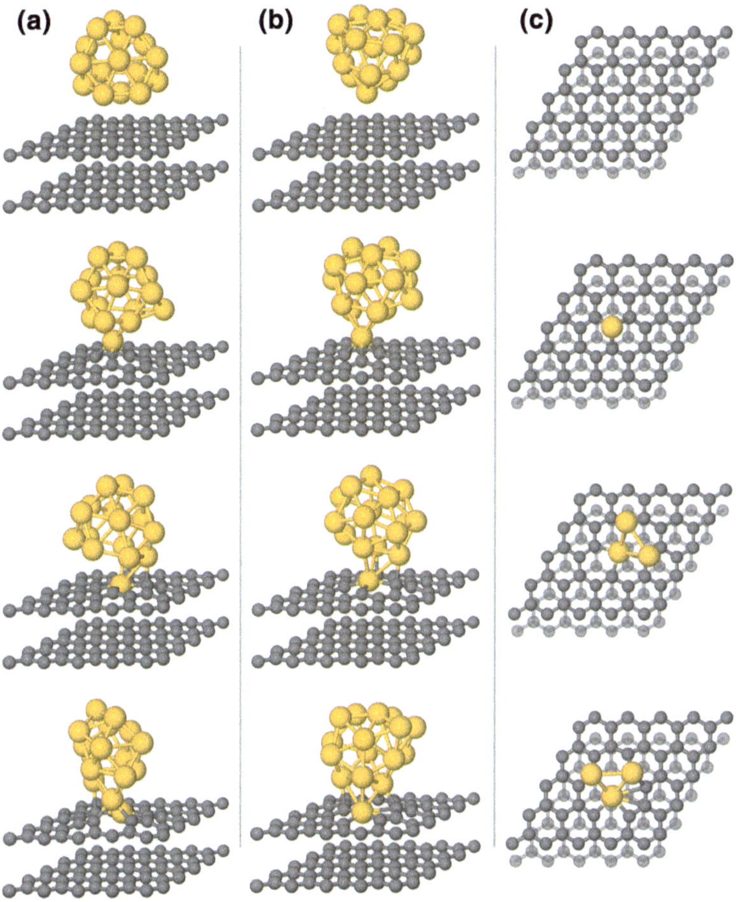

Fig. 4.7 (**a**) *Top* to *Bottom* Au_{16} structure absorbed on the (0001) face on to GR2, GR2-h, GR2-2h and GR2-3h, respectively. **b** As for (**a**), for cluster absorbed on the (111) face. **c** "Birds-eye" view of (**b**), with Au atoms > 3 Å from C atoms removed. Here the second layer of graphene has been made translucent to ease identification. In all images HOPG atoms were constrained beyond 3.5 Å of the cluster, and for the purpose of illustration some extraneous C atoms have been removed. Reprinted with permission from Ref. [30]. Copyright 2011 American Chemical Society

of adsorption with respect to geometrical changes in the cluster using ΔE, where positive values are energetically favourable.

We observe similar trends in the structure changes for adsorption on the Au_{16} (111) and (0001) faces: For GR2, the T_d-symmetry of the starting geometry is preserved, and the cluster energy relative to that of the isolated starting Au_{16} structure remains similar ($\Delta E \approx 0$ eV, Fig. 4.7). Calculations using the vdW-DF and vdW-DF2 functionals for (111) face adsorption on GR2 give increased $\bar{r}_{\text{Au-Au}}$ of 2.89 and 2.93 Å, respectively, due to expansion of the cluster under the increased van der Waals influence; the centre of mass for the Au_{16} cluster is not altered significantly.

For GR2-h the cluster's structure is distorted, with lowered symmetry (C_1), although the structure is energetically favourable for Au_{16} (111) face adsorption compared to that of the starting structure ($\Delta E = 0.09$ eV).

For GR2-2h, a gross distortion of the starting geometry occurs as an Au atom relocates from the cluster base into the double surface vacancy, contributing to decreased Au_{16} cluster energetic stability, which is seen most noticeably for Au_{16} (0001) face adsorption where $\Delta E = -1.43$ eV (Fig. 4.5a); this considerably affects the E_a of the interaction, as well as decreasing $N_{\text{Au–Au}}$ for the Au_{16} cluster. In the case of Au_{16} (111) face adsorption, $\bar{r}_{\text{Au–Au}}$ increases to 2.85 Å. The symmetry is low for these systems [C_1 and C_s for Au_{16} (111) and (0001) face adsorption, respectively] as the cage-like structure has been "opened up" with the displacement of the surface interacting Au atom from the cluster into the HOPG plane, while the inner void is preserved. The "tripod" positioning of a second Au atom on the HOPG surface for GR2-3h means that ΔE(GR2-3h) > ΔE(GR2-2h) [$\Delta E = -0.77$ and -0.20 eV for Au_{16} (111) and (0001) face adsorption on GR2-3h, respectively], which couples with increased E_a (Table 4.1).

For all adsorbed Au_{16} structures, $N_{\text{Au–Au}}$ remains similar [42 ± 1, with the exception of Au_{16} (0001) face adsorption on GR2-2h], showing quantitatively that the cage-like structure is maintained; should the structure have contracted inwards then $N_{\text{Au–Au}}$ would be expected to increase. For the exceptional case identified (GR2-2h), the strong adsorption of an Au atom into the HOPG plane leads to weaker long-range ("aurophilic" [53]) interactions with the rest of the cluster, and thus lower $N_{\text{Au–Au}}$ of 38. $\bar{r}_{\text{Au–Au}}$ increases with two or more surface defects, especially for Au_{16} (111) face adsorption on GR2-2h and GR2-3h (2.85 and 2.84 Å, respectively), due to dislocation of an Au atom from the cluster into the HOPG surface resulting in longer and weaker interactions to the rest of the cluster. T_d-symmetry is not preserved once a surface defect is introduced (GR2-h, GR2-2h and GR2-3h) despite the preservation of the cage-like features.

Structural observations of the adsorbed Au_{16} cluster with HOPG atoms constrained beyond 7.0 Å from the cluster are shown in the Appendix C.5. Observed trends for Au_{16} (111) and (0001) face adsorption are similar to those reported for a 3.5 Å radius of unconstrained atoms: $\bar{r}_{\text{Au–Au}}$ increases, and ΔE and symmetry decrease, with greater neighbouring surface defects (GR2 → GR2-3h). ΔE is lowest for GR2-2h for both Au_{16} (111) and (0001) face adsorption, as seen previously for 3.5 Å. Overall, changing the area of constrained HOPG atoms from 3.5 to 7.0 Å from the starting Au_{16} cluster shows minimal effect on structural properties. There are some differences in the ΔE values, but these could be attributed to greater influence of the C atoms on the Au_{16} orientations on the HOPG layers: a direct result of relaxing constraints.

4.4.3 Electronic Properties

PLDOS (Fig. 4.8) and decomposed orbital angular contributions to the PLDOS (Fig. 4.9) for Au_{16} adsorbed on GR2 on the (0001) face are shown. Calculations were performed with 13 special **k**-points to give better sampling of the HOPG surface, of which a radius of 3.5 Å of C atoms from the Au_{16} cluster were unconstrained. The HOPG DOS combined with a (0001) adsorbed Au_{16} in Fig. 4.8 shows typical features: a steep rise at −20 eV due to the 2D character of graphite, at −13 eV a dip after the first two σ bands, at −6.5 eV a large peak followed by a shoulder, and a zero gap and zero weight minimum close to the Fermi energy (E_f) [2, 49]. The charge transfer to Au_{16} has shifted slightly the location of the E_f deeper in the valence band. Despite the documented charge transfer, electron redistribution and atomic replacement of removed C atoms with an Au atom for a defect of size greater than one atom, the characteristic graphite features remain in each case, with exception of

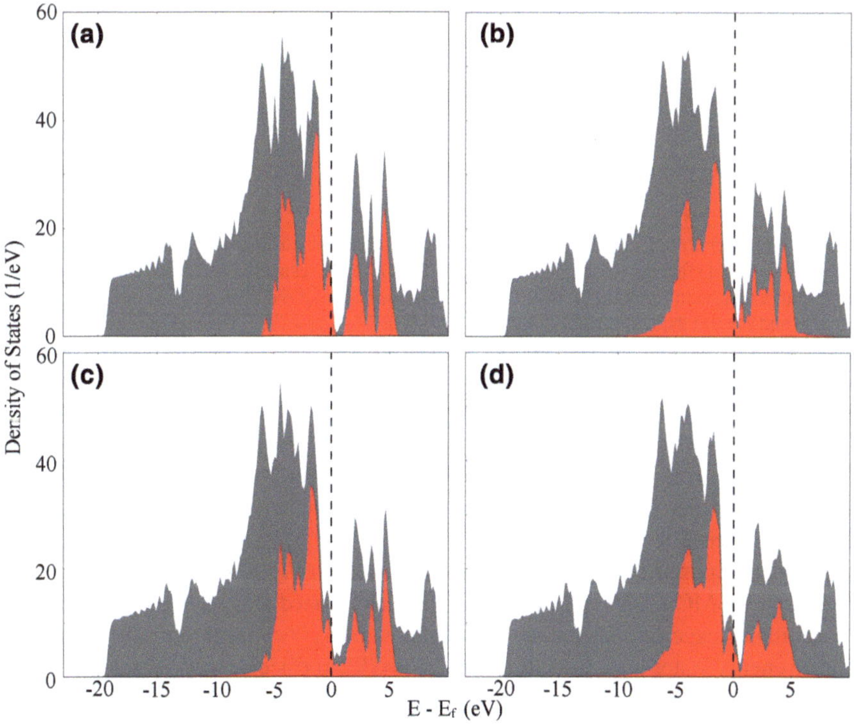

Fig. 4.8 From *top left* PLDOS for Au_{16} cluster (*red*) adsorbed on (0001) face on **a** GR2, **b** GR2-h, **c** GR2-3h and **d** GR2-2h, respectively (*grey*). Calculations were performed on a 5 × 5 × 1 Monkhorst-Pack grid, giving 13 special **k**-points, with points in the DOS fitted with a Gaussian function of width 0.2 eV. A *dashed vertical line* marks E_f. Reprinted with permission from Ref. [30]. Copyright 2011 American Chemical Society

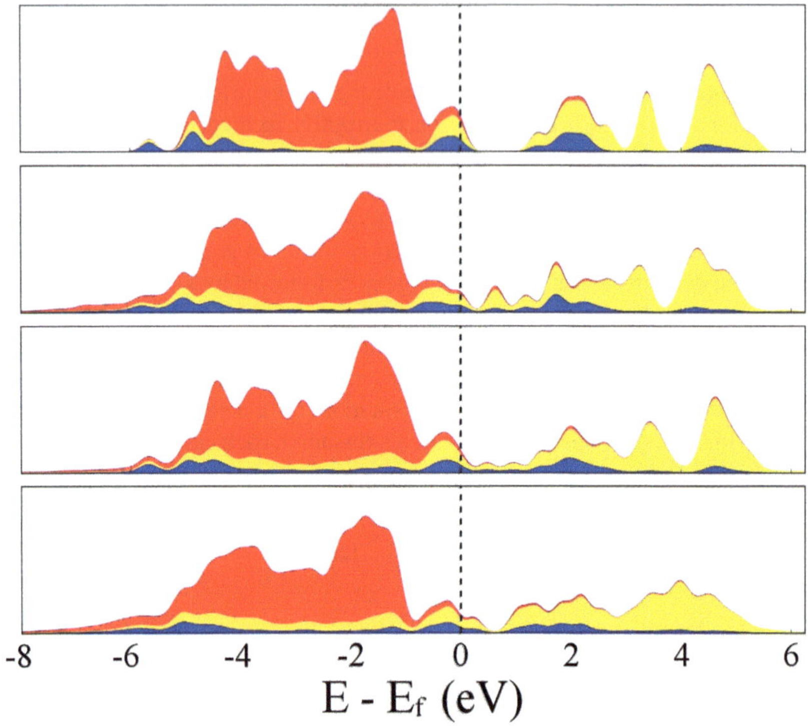

Fig. 4.9 *Top* to *bottom* Decomposed PLDOS, using valence electron projections, for Au_{16} cluster adsorbed on (0001) face on GR2, GR2-h, GR2-2h and GR2-3h, respectively. Points in the DOS were fitted with a Gaussian function of width 0.2 eV. *Blue*, *yellow* and *red* represent the contributions of the s, p and d orbitals, respectively. A *dashed vertical line* marks E_f. Reprinted with permission from Ref. [30]. Copyright 2011 American Chemical Society

the shape and positioning of the traditional "V-shaped" profile of HOPG around E_f. This is directly related to the changing electronic structure of the Au_{16} cluster as it is adsorbed further in-plane, which gives rise to a greater number of energy states close to the Fermi level.

The decomposed PLDOS for the Au_{16} clusters in Fig. 4.9 shows the closure of the cluster LUMO-LUMO+1 gap, previously mentioned as a source of stability for the (anionic) cage-like structure, with both s and p orbital contributions at E_f. Higher energy states appear as the number of GR2 surface defects increases. The d orbital contributions do not change their positioning considerably with the changes in structures. The electronic shell closing (band gap) remains evident for Au_{16} on GR2, supporting our previous observation that the cage-like structure is most stable when no defects are present.

4.4.4 Catalytic Activity

Small Au clusters have been used commercially for the catalysis of CO oxidation for many years [15, 16], and recently size-selected Au clusters supported on graphene have been studied computationally [12, 17]. To check the potential catalytic activity of deposited clusters on GR2 and defected GR2, a systematic search of adsorption sites for an O_2 molecule on all adsorbed Au_{16} systems was conducted, as is the case for the Langmuir-Hinshelwood mechanism of CO oxidation. The O_2 dimer was soft-landed on (a) the Au atom farthest vertically from the HOPG plane and (b) the Au atom farthest displaced laterally from the centre of mass for Au_{16} clusters adsorbed on both the (111) and (0001) faces. Calculations were performed with spin-unpaired electron configuration, allowing for triplet as well as singlet states. Binding energy (E_b) is used for analysis along with E_a; E_b is defined similarly to E_a as:

$$E_b = E^s_{ads} + E^s_{sub} - E_{tot} \tag{4.29}$$

but here E^s_{ads} and E^s_{sub} are the energies of the adsorbate (oxygen) and substrate (Au_{16} adsorbed on HOPG), respectively, in geometries isolated from the combined O_2 adsorbed on Au_{16}/HOPG system (i.e. without any further structural minimisation). This gives us a measure of the energy interaction between the adsorbed cluster system and dioxygen molecule (i.e. the strength of the Au–O bond) without considering structural changes of the isolated systems, as for E_a.

Gas-phase calculations for the O_2 minima give a triplet state with $E_b = 5.17$ eV, dimer bond length [d(O–O)] of 1.23 Å and a triplet-singlet energy difference $\Delta E = 1.12$ eV, which is similar to previous work [16, 54]. It is noted that the energetic minimum is a triplet in the gas-phase, however when successfully adsorbed the system switches to a singlet state.

For site type (a), it is noted that in general the O_2 molecule binds weakly to the Au surface ($E_b < 0.14$ eV), and adsorption is energetically unfavourable ($E_a < 0$), with $|E_a| > E_b$ implying that binding is unfavourable overall. Consequently, the O_2 bond length remained constant in the range 1.29 Å $\leq$ d(O–O) $\leq$ 1.30 Å. One exception was found for Au_{16} (111) adsorption on GR2-2h, where $E_a = 0.08$ eV, $E_b = 1.05$ eV and d(O–O) = 1.36 Å. This anomaly is due to the O_2 dimer positioning itself parallel to the cluster allowing more than one Au atom to interact with the molecule, and this causes distortion to the Au_{16} cluster (Fig. 4.10a). In this case Δq of the O_2 dimer is found to be 0.65 e, i.e. the molecule is in the superoxo-state. For laterally displaced sites [type (b)] the situation was as seen for type (a): $E_b \leq 0.18$ eV and $E_a < 0$, with the inequality $|E_a| > E_b$ holding true throughout, and d(O–O) was slightly less than for site (a), with a maximum of 1.28 Å.

For Au_{16} adsorbed on the (0001) face on GR2-2h a placement of the oxygen dimer within the "open cavity" was also attempted (Fig. 4.5). However, it was found that the dioxygen molecule readily migrates to the exterior of the Au_{16} cluster during geometry optimisation. Our findings are overall consistent with those recently presented by Zhou et al. for unstrained HOPG surfaces [12], and in most cases, they do

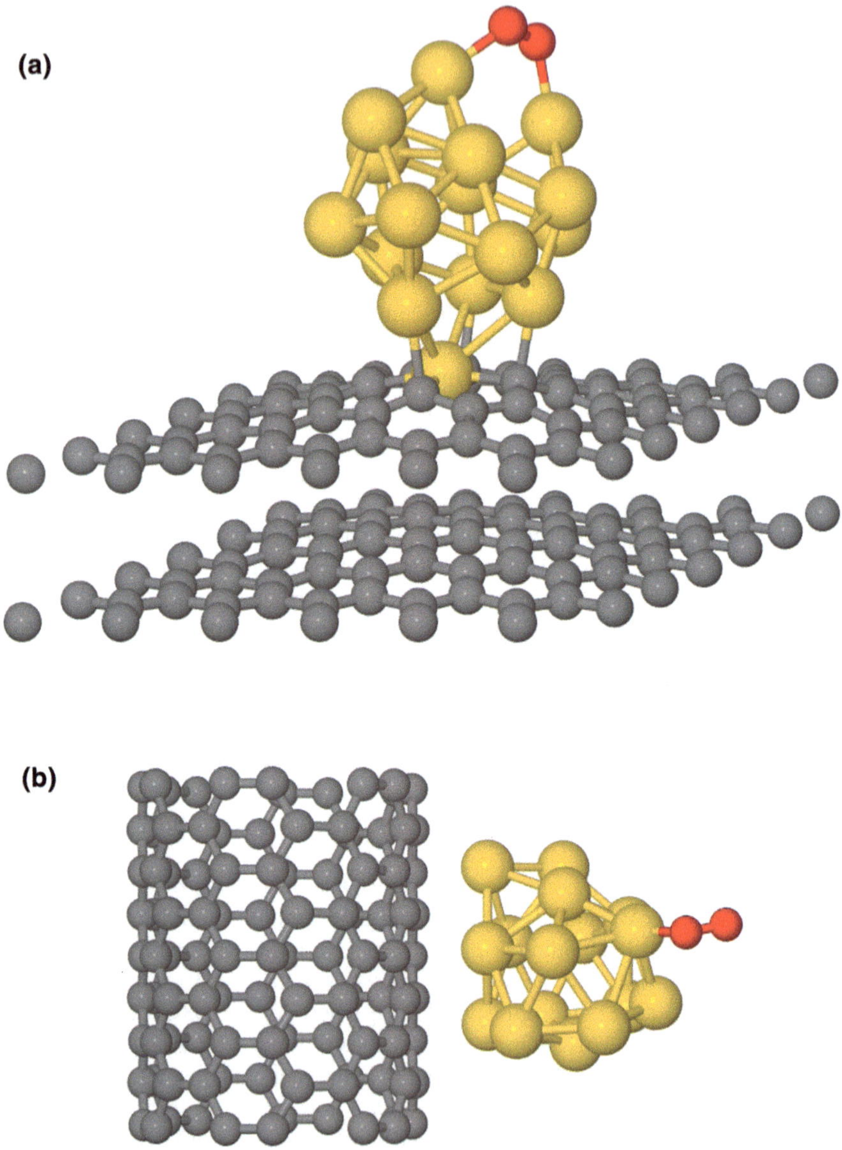

Fig. 4.10 O_2 dimer (*red*) adsorbed on to Au_{16} cluster on different HOPG surfaces: **a** O_2 dimer positioned between two Au atoms for Au_{16} (111) adsorption on GR2-2h, leading to elongated d(O–O) value. The distance between the O separated Au atoms, originally 2.71 Å, is now stretched to 3.45 Å to accommodate the O_2 adsorption. **b** Au_{16} cluster adsorbed on to a (6, 6) CNT. The decreased CNT-cluster interaction area is clear to see, and held responsible for the inability to catalytically activate the O_2 dimer. Reprinted with permission from Ref. [30]. Copyright 2011 American Chemical Society

not imply improved catalytic activity for Au_{16} in the presence of defects. Here, the single case of O_2 activation in Fig. 4.10a must be emphasised as a potential system for further reaction modelling; O_2 adsorption across two gold atoms has since been reported for an Au_{16} cluster deposited on graphane [55].

Of interest in Ref. [12] is the novel observation that induced strain in a perfect graphene sheet could lead to decreased Au–C distances and catalytic activation. We looked, therefore, at the adsorption of Au_{16} on to CNTs—which have strain present due to their curved nature—with a variation of diameters (8–14 Å) and chirality (zig-zag, armchair and chiral). Initial adsorption of the cluster proved promising: our best result was for an armchair CNT of ratio (6, 6), with 216 atoms and a diameter of 8.14 Å, giving E_a and Au-X of 0.99 eV and 2.99 Å, respectively. This is a considerable improvement on the pure GR2 surface. However, we could not activate the O_2 dimer: an adsorbed d(O–O) value of 1.27 Å was recorded (Fig. 4.10b), and there was no charge transfer into the dimer. Δq to the cluster is significantly less than on the HOPG surface (0.35 e), due to the lower contact area between the cluster and carbon surface, and so this is held responsible.

In comparison a CNT of ratio (15, 0), with 300 atoms and diameter 11.74 Å, gave E_a and Au-X of 0.53 eV and 2.94 Å, respectively, and Δq to the cluster of 0.71 e, which is significantly higher than the (6, 6) CNT, illustrating the importance of a high contact area for electron transfer. As for the HOPG surface, the cluster does not transfer the excess charge onto adsorbed O_2. Also, calculations for a chiral CNT of ratio (15, 6), with 468 atoms and a diameter of 14.67 Å, gave E_a and Au-X of 0.32 eV and 3.21 Å, respectively, much lower binding than for both previous CNT calculations, indicating that perhaps there is an optimal diameter (and chirality) CNT for Au_{16} adsorption within the range studied.

4.5 Adsorption on Doped HOPG Systems

HOPG substrate systems can be adapted to suit the needs of the adsorbate by the substitution or addition of electron-donating and electron-withdrawing elements. The electronegative characteristics of Au have been discussed in Sect. 4.4, and thus suitable substitution of nitrogen for a carbon atom, or addition of an alkali metal, would increase the density of electrons in the HOPG planes [56, 57], and thus aid Au adsorption.

Preliminary studies replacing one central surface C in the 144 C-atom supercell with a N atom were performed, with improvements in adsorption seen in previous cases [7]. As with our preceding calculations, C atoms beyond 3.5 Å from the Au_{16} cluster were fixed in their locations.

We found in our calculations that E_a did not change significantly for Au_{16} adsorption on to GR2-N on the 0001 face of the cluster (0.77 eV) and decreased slightly when adsorbed on the 111 face (0.59 eV) when compared the pure E_a on GR2 (0.78 and 0.80 eV on 0001 and 111 faces, respectively. See Table 4.1). The bond distance Au–N did not change greatly compared to Au–C in the GR2 systems (3.40 and

3.61 Å on the 111 and 0001 faces, respectively), nor did $d_{\perp}$(Au): 3.40 and 3.56 Å on the 111 and 0001 faces, respectively. Bader charge analysis shows that Δq for the N atoms is high in both cases (>1.2 e), with much of this charge being drawn from the bottom carbon layer. Δq for the Au_{16} cluster increases slightly to 0.79 e, in both cases, and so further investigation is necessary to see if potential catalytic activity can be improved.

Tentative preliminary calculations have also been performed for the intercalation of alkali metals M = Li, Na and K into the GR2 system. Restrictions were maintained on C atoms further than 3.5 Å from the Au_{16} cluster. Initial calculations appear promising, with stoichiometries of MC_{144} and MC_{16} (M = Li, Na) both offering positive E_a for the adsorption of the Au_{16} cluster, and encouraging positive values for Δq.

However, these initial intercalated studies have been conducted with an interlayer separation identical to that of GR2, fixed at the periphery to 3.35 Å, and previous work shows the graphite interlayer gap expands to 3.48–3.70 Å [58, 59] for LiC_6, 3.58 Å for LiC_8 and 4.78 Å for NaC_8 [57], as well as the AB stacking of HOPG transforming to AA stacking [57–59]. Therefore, in future work, it will be necessary to accommodate these factors into the calculations in order to give more physically realistic results.

4.6 Conclusions

We have studied the adsorption of a T_d-symmetry Au_{16} cluster on to two layers of HOPG, with varying numbers of defects and different cluster orientations. Our results show that the adsorption energy E_a increases with the number of surface defects, but this leads to heavy distortion of the starting structure due to the formation of chemical bonds between the Au_{16} cluster and defect site. These observed trends are in good agreement with previous work in this area [7, 17], illustrating trends which span small cluster sizes, the number of graphene layers and the number of surface defects.

Orientation of the deposited cluster is also found to have an effect. Throughout all calculations a degree of the starting cage-like feature remains in the structure. We find that as the Au_{16} cluster is able to move closer to the HOPG surface, with increasing surface defects, so the electronegativity of the isolated structure vanishes. Au atoms strongly prefer to be oriented in-plane, if possible, with the ideal vacancy size being that of two C-atoms (GR2-2h), as previously seen for single Au adatoms [11]. This is detrimental to cluster stability, as the attractive forces are so strong that they distort the structure, and potential catalytic activity for CO oxidation (activation of O_2), as $\Delta q \rightarrow 0$ for the cluster. Stability is best achieved (i.e. greatest E_a) on GR2-3h where a second Au atom can form "tripod"-like interactions over an in-plane Au and two C atoms [equivalent to a (111) surface site] and the out-of-plane Au cluster atoms. ELF illustrations show that the Au–C bond is delocalised ("metallic") in nature, unlike the covalent C–C bonds. The binding of an Au_{16} cluster to a defected substrate results in an increased bending of the substrate surface.

Additional test calculations for vdW-corrected exchange-correlation functionals lead to greater E_a and slightly shorter Au-X distances for cluster deposition on GR2, but this effect is reversed for GR2-2h due to the importance put on energy from long-distance van der Waals interactions, and the overall picture of the cluster-substrate interaction is unchanged. A conclusion as to whether these corrections are accurate for this particular system awaits further investigation.

The potential catalytic activity of Au_{16} clusters on graphitic supports has been probed in several cases. The T_d-symmetry and nearly complete electronic shell closing of Au_{16} mean that the cluster attracts electron density from HOPG and CNTs. However, the energetic interplay between electronic shell closing and O_2 adsorption results in the cluster not readily transferring the excess charge to O_2. For defects, it is noted that O_2 binding to the adsorbed Au_{16} clusters is unfavourable in most cases due to the depleted electron density at the perimeter of the cluster. However, there is one interesting exception for GR2-2h where the binding of the oxygen dimer distorts the underlying cluster geometry significantly, and the "pinched" O_2 is activated in the superoxo-state. Therefore, we cannot rule out the potential activity of these systems for CO oxidation.

Our observations are in agreement with those of Chen et al. [28], who postulated in their conclusions that transformation of the T_d-symmetry Au_{16} structure requires additional assistance to breach energetic barriers. In our results, the Au_{16} geometries are not as thermodynamically preferable as the starting isolated clusters, but they form strong and stable adsorbed arrangements. Further work is necessary to identify adsorption pathways to more thermodynamically favourable minima, as well as calculations to look at the effects of increased defect size and impact energy on cluster structure and adsorbed properties. Continued calculations, and analysis, of cluster deposition on doped-HOPG systems is also necessary to clarify their suitability as substrates for clusters in CO oxidation.

References

1. H. Haberland, *Clusters of Atoms and Molecules* (Springer, Berlin, 1994)
2. K. Rytkönen, J. Akola, M. Manninen, Sodium atoms and clusters on graphite: a density functional study. Phys. Rev. B **69**, 205404 (2004)
3. W. Hong, H. Bai, Y. Xu, Z. Yao, Z. Gu, G. Shi, Preparation of gold nanoparticle/graphene composites with controlled weight contents and their application in biosensors. J. Phys. Chem. C **114**(4), 1822–1826 (2010)
4. R. Muszynski, B. Seger, P.V. Kamat, Decorating graphene sheets with gold nanoparticles. J. Phys. Chem. C **112**(14), 5263–5266 (2008)
5. M. Di Vece, S. Palomba, R.E. Palmer, Pinning of size-selected gold and nickel nanoclusters on graphite. Phys. Rev. B **72**(7), 073407 (2005)
6. K. Kong, Y. Choi, B.H. Ryu, J.O. Lee, H. Chang, Investigation of metal/carbon-related materials for fuel cell applications by electronic structure calculations. Mater. Sci. Eng. C **26**(5–7), 1207–1210 (2006). Current trends in nanoscience: from materials to applications

7. J. Akola, H. Häkkinen, Density functional study of gold atoms and clusters on a graphite (0001) surface with defects. Phys. Rev. B **74**, 165404 (2006)
8. R.E. Palmer, S. Pratontep, H.G. Boyen, Nanostructured surfaces from size-selected clusters. Nature Mater. **2**, 443 (2003)
9. D.C. Lim, R. Dietsche, M. Bubek, T. Ketterer, G. Ganteför, Y.D. Kim, Chemistry of mass-selected Au clusters deposited on sputter-damaged HOPG surfaces: the unique properties of Au_8 clusters. Chem. Phys. Lett. **439**(4–6), 364–368 (2007)
10. Y. Gan, L. Sun, F. Banhart, One- and two-dimensional diffusion of metal atoms in graphene. Small **4**, 587–591 (2008)
11. S. Malola, H. Häkkinen, P. Koskinen, Gold in graphene: in-plane adsorption and diffusion. Appl. Phys. Lett. **94**, 043106 (2009)
12. M. Zhou, A. Zhang, Z. Dai, Y.P. Feng, C. Zhang, Strain-enhanced stabilization and catalytic activity of metal nanoclusters on graphene. J. Phys. Chem. C **114**(39), 16541–16546 (2010)
13. J. Akola, M. Walter, R.L. Whetten, H. Häkkinen, H. Grönbeck, On the structure of thiolate-protected Au_{25}. J. Am. Chem. Soc. **130**(12), 3756–3757 (1999)
14. M. Walter, J. Akola, O. Lopez-Acevedo, P.D. Jadzinsky, G. Calero, C.J. Ackerson, R.L. Whetten, H. Grönbeck, H. Häkkinen, A unified view of ligand-protected gold clusters as superatom complexes. Proc. Natl. Acad. Sci. **105**, 9157–9162 (2008)
15. M. Haruta, S. Tsubota, T. Kobayashi, H. Kageyama, M.J. Genet, B. Delmon, Low-temperature oxidation of CO over gold supported on TiO_2, α-Fe_2O_3, and Co_3O_4. J. Catal. **144**(1), 175–192 (1993)
16. O. Lopez-Acevedo, K.A. Kacprzak, J. Akola, H. Häkkinen, Quantum size effects in ambient CO oxidation catalysed by ligand-protected gold clusters. Nature Chem. **2**, 329–334 (2010)
17. J.P. Liu, E. Fullerton, O. Gutfleisch, D.J. Sellmyer, *Nanoscale Magnetic Materials and Applications* (Springer, New York, 2009)
18. J. Wang, G. Wang, J. Zhao, Density-functional study of $Au_n (n = 2–20)$ clusters: lowest-energy structures and electronic properties. Phys. Rev. B **66**(3), 35418–35424 (2002)
19. X. Xing, B. Yoon, U. Landman, J.H. Parks, Structural evolution of Au nanoclusters: from planar to cage to tubular motifs. Phys. Rev. B **74**(16), 165423 (2006)
20. G. Santarossa, A. Vargas, M. Iannuzzi, A. Baiker, Free energy surface of two- and three-dimensional transitions of Au_{12} nanoclusters obtained by ab initio metadynamics. Phys. Rev. B **81**(17), 174205 (2010)
21. M.P. Johansson, A. Lechtken, D. Schooss, M.M. Kappes, F. Furche, 2D–3D transition of gold cluster anions resolved. Phys. Rev. A **77**(5), 053202 (2008)
22. J. Li, X. Li, H.J. Zhai, L.S. Wang, Au_{20}: a tetrahedral cluster. Science **299**(5608), 864–867 (2003)
23. P. Gruene, D.M. Rayner, B. Redlich, A.F.G. van der Meer, J.T. Lyon, G. Meijer, A. Fielicke, Structures of neutral Au_7, Au_{19}, and Au_{20} clusters in the gas phase. Science **321**(5889), 674–676 (2008)
24. S. Bulusu, X. Li, L.S. Wang, X.C. Zeng, Evidence of hollow golden cages. Proc. Natl. Acad. Sci. **103**(22), 8326–8330 (2006)
25. M. Walter, H. Häkkinen, A hollow tetrahedral cage of hexadecagold dianion provides a robust backbone for a tuneable sub-nanometer oxidation and reduction agent via endohedral doping. Phys. Chem. Chem. Phys. **8**, 5407 (2006)
26. L.M. Wang, R. Pal, W. Huang, X.C. Zeng, L.S. Wang, Tuning the electronic properties of the golden buckyball by endohedral doping: $M@Au_{16}^-$ (M = Ag, Zn, In). J. Chem. Phys **130**(5), 051101 (2009)
27. L.M. Wang, S. Bulusu, W. Huang, R. Pal, L.S. Wang, X.C. Zeng, Doping the golden cage Au_{16}^- with Si, Ge, and Sn. J. Am. Chem. Soc. **129**(49), 15136–15137 (2007)
28. G. Chen, Q. Wang, Q. Sun, Y. Kawazoe, P. Jena, Structures of neutral and anionic Au_{16} clusters revisited. J. Chem. Phys. **132**(19), 194306 (2010)
29. W.A. de Heer, The physics of simple metal clusters: experimental aspects and simple models. Rev. Mod. Phys. **65**(3), 611–676 (1993)

30. A.J. Logsdail, J. Akola, Interaction of Au_{16} nanocluster with defects in supporting graphite: a density-functional study. J. Phys. Chem. C **115**(31), 15240–15250 (2011)
31. F. Jensen, *Introduction to Computational Chemistry* (Wiley, Chichester, 1999)
32. R.G. Parr, W. Yang, *Density Functional Theory of Atoms and Molecules* (New York, Oxford, 1989)
33. P. Hohenberg, W. Kohn, Inhomogeneous electron gas. Phys. Rev. **136**, B864–B871 (1964)
34. W. Kohn, L.J. Sham, Self-consistent equations including exchange and correlation effects. Phys. Rev. **140**, A1133–A1138 (1965)
35. A.J. Cohen, P. Mori-Sánchez, W. Yang, Challenges for density functional theory. Chem. Rev. **112**(1), 289–320 (2012)
36. A. Hinchliffe, *Chemical Modelling: From Atoms to Liquids* (Wiley, Chichester, 1999)
37. P.E. Blöchl, Projector augmented-wave method. Phys. Rev. B **50**(24), 17953–17979 (1994)
38. J. Enkovaara, C. Rostgaard, J.J. Mortensen, J. Chen, M. Dulak, L. Ferrighi, J. Gavnholt, C. Glinsvad, V. Haikola, H.A. Hansen, H.H. Kristoffersen, M. Kuisma, A.H. Larsen, L. Lehtovaara, M. Ljungberg, O. Lopez-Acevedo, P.G. Moses, J. Ojanen, T. Olsen, V. Petzold, N.A. Romero, J. Stausholm-Møller, M. Strange, G.A. Tritsaris, M. Vanin, M. Walter, B. Hammer, H. Häkkinen, G.K.H. Madsen, R.M. Nieminen, J.K. Nørskov, M. Puska, T.T. Rantala, J. Schiøtz, K.S. Thygesen, K.W. Jacobsen, Electronic structure calculations with gpaw: a real-space implementation of the projector augmented-wave method. J. Phys.: Condens. Matter **22**(25), 253202 (2010)
39. J.J. Mortensen, L.B. Hansen, K.W. Jacobsen, Real-space grid implementation of the projector augmented wave method. Phys. Rev. B **71**(3), 035109 (2005)
40. J.P. Perdew, K. Burke, M. Ernzerhof, Generalized gradient approximation made simple. Phys. Rev. Lett. **77**(18), 3865–3868 (1996)
41. E.R. Davidson, The iterative calculation of a few of the lowest eigenvalues and corresponding eigenvectors of large real-symmetric matrices. J. Comp. Phys. **17**, 87–94 (1975)
42. B.T. Kelly, *Physics of Graphite* (Applied Science Publishers, London, 1981)
43. K.R. Kganyago, P.E. Ngoepe, Structural and electronic properties of lithium intercalated graphite LiC_6. Phys. Rev. B **68**(20), 205111 (2003)
44. M. Dion, H. Rydberg, E. Schröder, D.C. Langreth, B.I. Lundqvist, Van der Waals density functional for general geometries. Phys. Rev. Lett. **92**(24), 246401 (2004)
45. K. Lee, É.D. Murray, L. Kong, B.I. Lundqvist, D.C. Langreth, Higher-accuracy van der Waals density functional. Phys. Rev. B **82**(8), 081101 (2010)
46. D.F. Shanno, Conditioning of quasi-Newton methods for function minimization. Math. Comp. **24**, 647–656 (1970)
47. W. Tang, E. Sanville, G. Henkelman, A grid-based Bader analysis algorithm without lattice bias. J. Phys.: Comput. Mater. **21**, 084204 (2009)
48. A. Savin, R. Nesper, S. Wengert, T.F. Fässler, Elf: the electron localization function. Angew. Chem. Int. Ed. **36**(17), 1808–1832 (1997)
49. K. Rytkönen, J. Akola, M. Manninen, Density functional study of alkali metal atoms and monolayers on graphite (0001). Phys. Rev. B **75**, 075401 (2007)
50. C. Kittel, *Introduction to Solid State Physics*, 6th edn. (Wiley, New York, 1986)
51. P. Jensen, X. Blase, P. Ordejon, First principles study of gold adsorption and diffusion on graphite. Surf. Sci. **564**(1–3), 173–178 (2004)
52. P.O. Lehtinen, A.S. Foster, Y. Ma, A.V. Krasheninnikov, R.M. Nieminen, Irradiation-induced magnetism in graphite: a density functional study. Phys. Rev. Lett. **93**(18), 187202 (2004)
53. F. Scherbaum, A. Grohmann, B. Huber, C. Krüger, H. Schmidbaur, "Aurophilicity" as a consequence of relativistic effects: the hexakis(triphenylphosphaneaurio)methane dication $[(Ph_3PAu)_6C]^{2+}$. Angew. Chem. Int. Ed. **27**(11), 1544–1546 (1988)
54. P. Giannozzi, R. Car, G. Scoles, Oxygen adsorption on graphite and nanotubes. J. Chem. Phys. **118**(3), 1003–1006 (2003)
55. G. Chen, S.J. Li, Y. Su, V. Wang, H. Mizuseki, Y. Kawazoe, Improved stability and catalytic properties of Au_{16} cluster supported on graphane. J. Phys. Chem. C **115**(41), 20168–20174 (2011)

56. Y. Ma, A.S. Foster, A.V. Krasheninnikov, R.M. Nieminen, Nitrogen in graphite and carbon nanotubes: magnetism and mobility. Phys. Rev. B **72**, 205416 (2005)
57. C. Hartwigsen, W. Witschel, E. Spohr, Charge density and charge transfer in stage-1 alkali-graphite intercalation compounds. Phys. Rev. B **55**, 4953–4959 (1997)
58. G. Sun, M. Kertesz, J. Kürti, R.H. Baughman, Dimensional change as a function of charge injection in graphite intercalation compounds: a density functional theory study. Phys. Rev. B **68**, 125411 (2003)
59. N.A.W. Holzwarth, S.G. Louie, S. Rabii, Lithium-intercalated graphite: self-consistent electronic structure for stages one, two, and three. Phys. Rev. B **28**, 1013–1025 (1983)

Chapter 5
Simulating the Optical Properties of Au Nanoclusters and Nanoalloys

5.1 Introduction

Plasmonics is the title given to the optical phenomenon related to plasma oscillations in metals. Original work on surface-plasmons is attributed to Ritchie [1], who postulated that electrons travelling through thin films would experience less energy loss than those travelling through bulk, which was subsequently verified by Stern and Ferrell [2]. The electromagnetic excitation of surface-plasmons at planar surfaces is called surface-plasmon resonance (SPR).

For discrete nanoparticles, the high surface to volume ratio dictates that surface effects have a much stronger influence on the optical properties than would be expected in bulk equivalents; surface-plasmons at this size regime are referred to as localised SPRs. Nanoplasmonics has received considerable attention in recent times due to technological advances which now allow the manipulation and structural characterisation of clusters on the nanometre scale [3], and the resultant applications of these features e.g. photovoltaic cells [4]. The optical properties of metallic nanoparticles prove to be size, shape and composition dependent, allowing for tunability; in the case of heterogeneous nanoclusters the complexity of composition contributions are further complicated by the different elemental species [5]. Furthermore, the optical properties can also be altered by the surrounding medium [6].

Theoretical calculations compliment experimental work in the field of optical characterisation, and allow for the identification of novel avenues of investigation. Exact solutions to Maxwell's equations of electromagnetism were first offered by Mie [7], and are only known for special geometries [8–10]. Therefore, approximate methods are often required; an example of which is the discrete-dipole approximation (DDA). The DDA is a flexible method for computing the absorption and scattering, components of the extinction, for nanoparticles with an arbitrary geometry [11], which works by simulating the nanoparticles as a defined array of polarisable points, which in turn acquire dipole moments as a result of interaction with a local electric field.

A. J. Logsdail, *Computational Characterisation of Gold Nanocluster Structures*,
Springer Theses, DOI: 10.1007/978-3-319-01493-7_5,

Optical absorption is strongly influenced by the electronic structure of particles, which is influenced by the factors outlined above. In general, nanoclusters of group XI elements (Cu, Ag, Au) are shown to have features in the ultraviolet and visible (UV-Visible) spectral range [12, 13], making them of specific interest for nano-optical applications. This absorption is associated with the promotion of a d band electron to the unoccupied upper s orbitals [14]. On the other hand, nanoclusters of group X elements (Pd, Pt) give featureless spectra in this wavelength region: $d \rightarrow s$ transitions prove to be too high in energy relative in intraband alternatives.

Gold is shown to absorb electromagnetic radiation in the visible region, with a maximum extinction at the wavelength (λ_{max}) $\approx$ 520 nm [15]. Geometry and size plays an important factor in the location of λ_{max}, and Au nanoparticles are of particular interest due to the tunability of their extinction spectra via variation of geometrical factors: experimental work, accompanied by DDA calculations, has compared results and characterised optical responses for Au nano-spheres [16], nano-rods [13, 17–19], nano-disks [20] and nano-cages [21], and also looked at coupling between some of these geometries [16, 22, 23]. Elongated structures, such as ellipsoids, cylinders and cuboids, are shown to cause λ_{max} to red-shift with increasing difference between the long- and short-axis geometric dimensions [24].

The susceptibility of λ_{max} to geometric factors proves consistent for other group XI elements: Cu and Ag have extinction peaks at ~600 and 350 nm in the visible spectrum for spherical particles, however elongation of one axis results in red-shifting of the prominent λ_{max} [25, 26].

Tuning λ_{max} can also be achieved by creating bimetallic structures, which are shown to offer novel properties compared to their constituent elements [27, 28]. Au-Pd clusters are of high industrial interest due to the catalytic properties of the system [29, 30]. Au-Pd clusters have been shown by previous work to be able to form both segregated arrangements of $Au_{core}Pd_{shell}$ and $Pd_{core}Au_{shell}$ [5, 31]. A Pd shell quenches the aforementioned Au optical features: reports by Toshima [32] and Liu et al. [29] on the effect of molar ratio for spherical Au-Pd clusters show that Au absorbance spectra are suppressed by even small ratios of Pd.

The Xiang group have reported reducing Pd onto Au rods producing rectangular shaped $Au_{core}Pd_{shell}$ nanorods [33]; this has been corroborated by Song et al., [34], who also studied $Au_{core}Ag_{shell}$ systems and concluded that thin-shells on an Au seed can have a strong influence on the optical extinction. $Au_{core}Ag_{shell}$ nanoparticles have been the subject of extended investigation, with spherical particles being particularly well studied [28, 35]; Li et al. have studied both spherical $Au_{core}Ag_{shell}$ nanoparticles and the inverse $Ag_{core}Au_{shell}$ structure, with thin-shells again proving influential to the evolution of extinction characteristics in this case [36, 37].

In this chapter, we look at the influence of geometry and chemical composition on the absorption spectra for Au nanoparticles, with comparisons to experiment along the way. We will start by defining the methodology, before looking at spherical and elliptical nanoparticles, and then finishing with some nanorod case studies.

5.2 Methodology

5.2.1 The Drude-Sommerfield Model

One of the simplest models for valence electron behaviour in macroscopic metal crystal structure is offered by the Drude-Sommerfield electron model, whereby the conduction electrons are treated as a homogeneous gas with the ability to oscillate about the fixed ionic core due to Coulombic attractive forces (Fig. 5.1). When an electric field is applied to the system, as in the presence of electromagnetic radiation (λ), electrostatic coupling between the electron cloud and the time-varying electric field at resonant frequencies causes a coherent displacement of electrons, which are then subjected to a restoring electrostatic force once the electromagnetic wave has passed. Energy is transferred between electrostatic (potential) and kinetic forms by the oscillating plasma particles, and a plasmon is defined as an energy quantum of these plasma oscillations. The plasmon oscillation frequency (ω_p) is defined as [6]:

$$\omega_p = \sqrt{\frac{ne^2}{m_e \varepsilon_0}} \tag{5.1}$$

where n is the number density of electrons, e is the charge of an electron, m_e is the mass of an electron and ε_0 is the permittivity of free space. For plasmon oscillations at the material boundary the generated electric fields extend into the surrounding dielectric medium, ε_m, forming what is known as surface plasmon resonances (SPRs) as shown in Fig. 5.2, with an angular frequency (ω_{SP}) defined as:

$$\omega_{SP} = \frac{\omega_p}{\sqrt{\varepsilon_m + 1}} \tag{5.2}$$

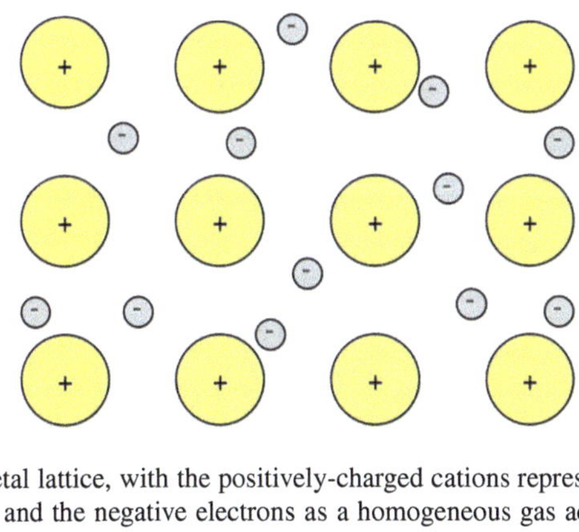

Fig. 5.1 An ionic metal lattice, with the positively-charged cations represented as fixed particles (*large yellow circles*) and the negative electrons as a homogeneous gas across the system (*small grey circles*)

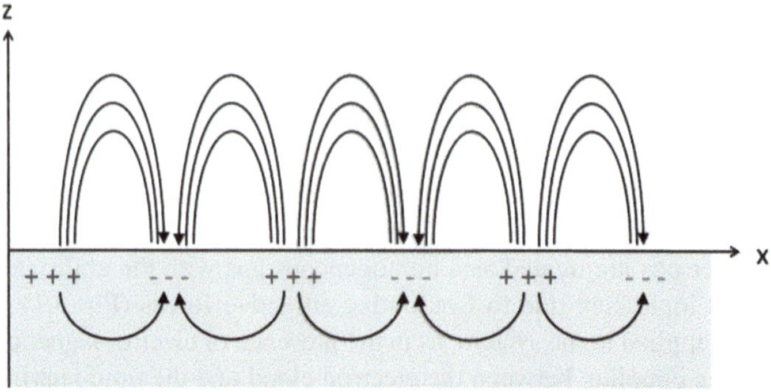

Fig. 5.2 Charge fluctuations at a dielectric interface, with field lines flowing from positive (+) to negative (−): Electric field lines propagate further into the dielectric medium (ε_m), represented with a *white* background, than into the metal, shown with a *grey* background

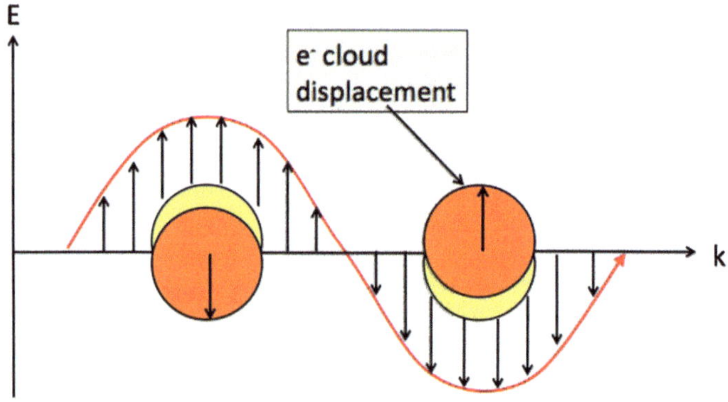

Fig. 5.3 Schematic diagram illustrating the displacement of the conduction electrons (*orange circle*) from an ionic core (*yellow circle*) under the influence of an oscillating electromagnetic field (*red line*). Axes are the direction of propagation, *k*, and the direction of the oscillating electric field, *E*

For nanoparticles the response to photons is almost entirely dominated by surface effects; distinction from bulk SPRs is necessary due to differences between the planar bulk surface and discrete nanoparticle surface, and so SPRs for nanoparticles are known as localised SPRs. Localised SPRs can be approximated as dipolar for sizes below 150 nm at visible wavelengths, and this is known as the Dipole Approximation (Fig. 5.3) [38]. For nanoparticles where all dimensions <1 % of λ, the near infinite speed of the electric field across the particle can be taken into consideration as what is known as the Quasi-static Approximation. For <2 nm particles the classical band structure is inadequate, due to the discrete energy levels, and quantum solutions must be sort [15, 39].

Wavelength dependent dielectric functions are experimental measurable for bulk surfaces (ε_{exp}); they are a superposition of both intraband (ε_{intra}) and interband (ε_{inter}) contributions such that:

$$\varepsilon_{exp}(\lambda) = \varepsilon_{inter}(\lambda) + \varepsilon_{intra}(\lambda) \tag{5.3}$$

where ε_{inter} contributions are a result of high energy electron transitions between occupied and unoccupied bulk bands, separated by an energy gap e.g. 5d → 6s transitions in Au. ε_{intra} contributions represent lower energy electron transitions close to the Fermi level in incompletely filled bands: electrons at the Fermi level in metals are excited by very low energy photons, and as such can be described as "free" electrons [24]. The dielectric functions provided in the literature are of the form:

$$\varepsilon_{exp}(\lambda) = \varepsilon_1(\lambda) + i\varepsilon_2(\lambda) = (n + i\kappa)^2 \tag{5.4}$$

where the λ dependent real (ε_1) and imaginary (ε_2) components of the dielectric function are associated to the storage and dissipation of energy within the medium, and can be directly related to the real (n) and imaginary (κ) components of the refractive index [23].

5.2.2 Mie Theory and the Quasi-Static Approximation

The first description of the optical properties of nanoparticles, within an electrodynamical framework, were calculated exactly by Gustav Mie for spherical particles [7]. The resultant scattering (C_{sca}) and extinction (C_{ext}) cross-section efficiencies are given by:

$$C_{sca} = \frac{2\pi r^2}{x^2}\sum_{n=1}^{\infty}(2n+1)(|a_n|^2 + |b_n|^2) \tag{5.5}$$

$$C_{ext} = \frac{2\pi r^2}{x^2}\sum_{n=1}^{\infty}(2n+1)\Re e(a_n + b_n) \tag{5.6}$$

in which the size parameter, x, is given by [23]:

$$x = \frac{2\pi r \varepsilon_m}{\lambda} \tag{5.7}$$

where r is the particle radius and a_n and b_n are the Mie scattering coefficients which may be described in terms of spherical Bessel functions [40]; n then corresponds to the mode of excitement, where dipole ($n = 1$), quadrupole ($n = 2$) and higher orders ($n > 2$) are calculable, though with decreasing influence at small sizes.

The scattering cross section (C_{sca}) can be obtained from the conservation relationship:

$$C_{ext} = C_{sca} + C_{abs} \tag{5.8}$$

For particles much smaller than the wavelength of light, i.e. in the quasi-static regime, the dipole approximation is equal to the first order Mie calculations ($n = 1$ in Eqs. 5.5 and 5.6). In the quasi-static regime scattering and absorption can be calculated as:

$$C_{abs} = k\Im m(\alpha) \tag{5.9}$$

$$C_{sca} = \frac{k^4}{6\pi}|\alpha|^2 \tag{5.10}$$

where k is the wave vector $(= 2\pi\sqrt{\varepsilon_m}/\lambda)$; $\varepsilon_m = 1$ is used for a vacuum. The polarisability of a sphere, α_{sph}, used in the above equations is given by the dipolar Clausius-Mossotti equation [6]:

$$\alpha_{sph} = 4\pi r^3\left(\frac{\varepsilon_{exp} - \varepsilon_m}{\varepsilon_{exp} + 2\varepsilon_m}\right) \tag{5.11}$$

5.2.2.1 Mie-Gans Theory for Ellipsoids

With few exceptions, such as the for an infinitely long cylinder [10], Mie theory cannot be used to calculate light scattering and absorption for non-spherical particles. However, within the quasi-static approximation introduction of a geometrical factor, as suggested by Gans, allows determination of responses for prolate and oblate ellipsoids [9, 41]. In several papers metallic nanorods are addressed as prolate ellipsoids; the polarisability is described by introducing a depolarisation factor $L_{x,y,z}$ into Eq. 5.11 [13]:

$$\alpha_{x,y,z} = \frac{4\pi}{3}abc\left[\frac{\varepsilon_{exp} - \varepsilon_m}{\varepsilon_m + L_{x,y,z}(\varepsilon_{exp} + 2\varepsilon_m)}\right] \tag{5.12}$$

with the radii a, b and c reflecting the particle length along the x, y and z axes. For ellipsoids, and other prolate geometric structures where the main axes are unequal (i.e. $a > b = c$), the aspect ratio (AR) is defined as a/b. $L_{x,y,z}$ denotes the depolarisation along the semi-axes in a cartesian frame, with increased displacement expected along the elongated axis (x) for ellipsoidal particles such that [24, 41]:

$$L_x = \frac{1 - e^2}{e^2}\left(\frac{1}{2e}\ln\frac{1+e}{1-e} - 1\right) \tag{5.13}$$

where $e = \sqrt{1 - (b/a)^2}$. Using the depolarisation sum rule ($\sum_\beta L_\beta = 1$, where $\beta = x, y, z$) we can show the influence of L_y and L_z therefore becomes:

$$L_y = L_z = \frac{1}{2}(1 - L_x) \tag{5.14}$$

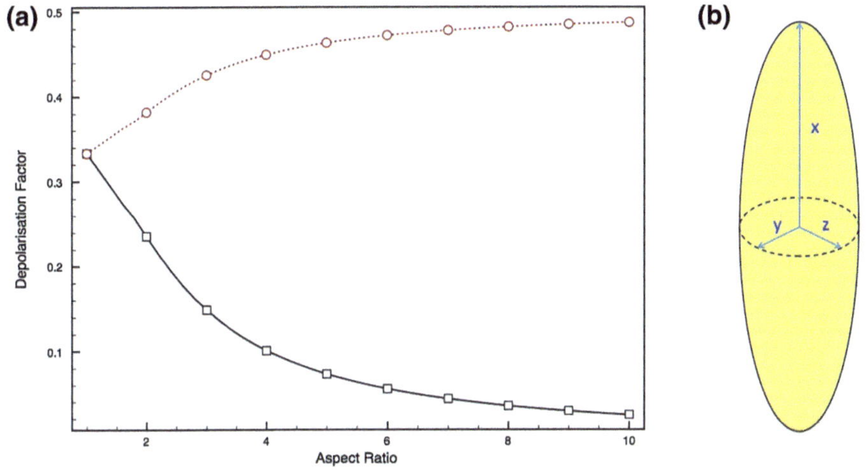

Fig. 5.4 **a** The depolarisation factors for the long axis, L_x (*black, solid line*), and short axes, $L_y = L_z$ (*red, dashed line*). When AR $= 1$, a sphere is regained as $L_{x,y,z} \rightarrow 1/3$. **b** The three axes of the prolate ellipsoid, where the length of the x-axis, a, is greater than the lengths of the y- and z-axis, b and c, respectively (i.e. $a > b = c$)

This is plotted in Fig. 5.4.

Gans' equation predicts that the plasmon peak mode varies with AR for small elliptical metal nanoparticles; this obvious conclusion stimulated much further work in the field, both experimentally and theoretically, with the implication clear that if the growth parameters can be controlled then materials with a tunable extinction spectra could be synthesised [41].

Excitation of the elliptical particles is orientation dependent, with k parallel to the elongated x-axes causing resonance of the transverse SPR (TSPR) and k perpendicular to the elongated x-axes (i.e. the electric field, E, parallel to the axes) resulting in the polarisation of the longitudinal SPR (LSPR), as illustrated in Fig. 5.5.

5.2.3 Heterogeneous Bimetallic Systems

Improved tunability of metallic nanoparticles is possible with the combination of more than one metal either by (a) coating a shell-layer of an alternative material, with dielectric ε_{shell}, on to a core with dielectric ε_{core} [41], to form a core-shell particle or (b) through alloying two materials together to provide novel electronic properties [6]. In the former case, the previously described dielectric polarisability of an ellipsoid ($\alpha_{x,y,z}$, Eq. 5.12) can be modified within the electrostatic limit to give [41]:

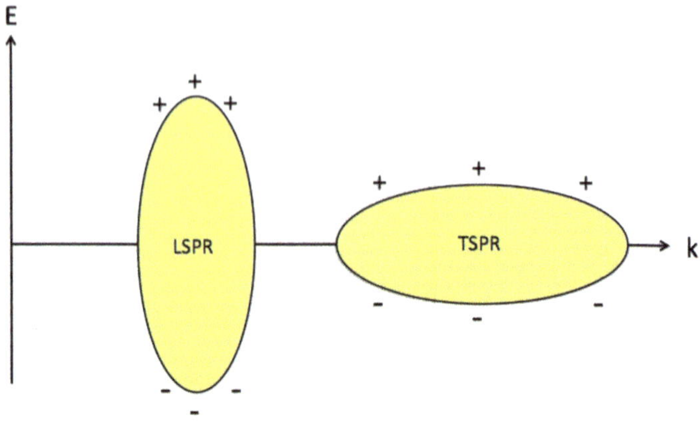

Fig. 5.5 Polarisation effects along the long and short axes of an ellipsoid, with *k* propagating across the page. *Left* x-axis aligned perpendicular to *k*, with instantaneous net charge displacements (+ and −), at separate ends of the ellipsoid forming a LSPR. *Right* x-axis aligned parallel to *k*, resulting in net charge displacements across the short (y- and z-) axes of the ellipsoid as found for a TSPR

$$\alpha_x = V\frac{(\varepsilon_{shell}-\varepsilon_m)[\varepsilon_{shell}+(\varepsilon_{core}-\varepsilon_{shell})(L^x_{core}-fL^x_{shell})]+f\varepsilon_{shell}(\varepsilon_{core}-\varepsilon_{shell})}{[\varepsilon_{shell}+(\varepsilon_{core}-\varepsilon_{shell})(L^x_{core}-fL^x_{shell})][\varepsilon_m+(\varepsilon_{core}-\varepsilon_m)L^x_{shell}]\varepsilon_{shell}(\varepsilon_{core}-\varepsilon_{shell})} \tag{5.15}$$

where the depolarisation factors L^x_{shell} and L^x_{core} are representative of the depolarisation factors for the core and shell, respectively; V is the volume of the entire particle ($= 4\pi a_{shell}b_{shell}c_{shell}/3$), and f is the volume fraction $a_{core}b_{core}c_{core}/a_{shell}b_{shell}c_{shell}$ occupied by the inner ellipsoid (Fig. 5.6). As expected, when $\varepsilon_{shell} = \varepsilon_{core}$ and $f = 1$, we recover Eq. 5.12.

Qualitatively, one would expect the extinction profile for a heterogenous composite structure to be a combination of the two individual spectra of the component metals. For instance, an Ag coated Au sphere would show multiple peaks, with the Ag features increasing in intensity and the Au peak decreasing in intensity as the shell thickens towards the skin depth, at which point the extinction profile reflects that of the shell material only [36]. The skin depth for Au and Ag is approximately 30 nm [39].

5.2.4 Modifications for Coupled Systems

The normal dipolar electric fields produced by an isolated nanoparticle extend out from the surface with a range reaching up to $\lambda/2$ in the surrounding medium, decaying exponentially over distance, where as the decay length within the metal is determined by the skin depth [43]. The fields extending out into the medium are termed "near fields" to reflect their short range influence. When two or more spheres are brought into close proximity the near fields can weakly couple the two nanoparticle

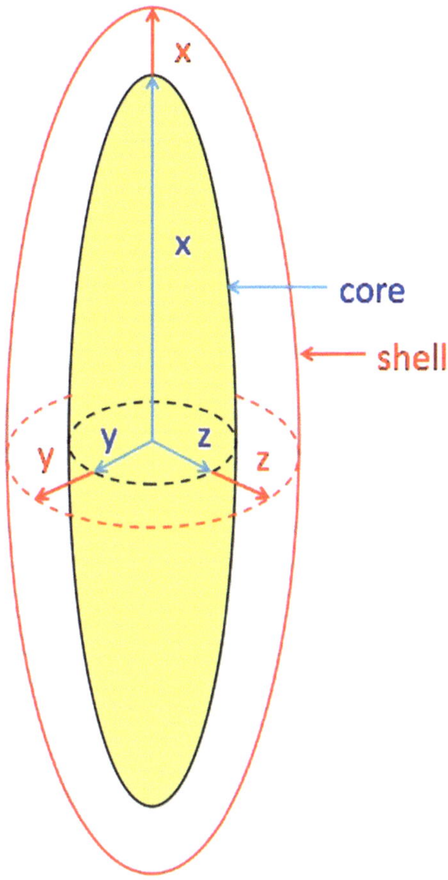

Fig. 5.6 The core-shell model of an ellipsoid, as adapted from Fig. 5.4b. The geometric parameters of the core are given in *blue*, whilst those of the shell are given in *red*. The volume of the entire system is found using the summated length for each dimension x-, y- and z-, respectively. Reprinted with permission from Ref. [42]. Copyright 2012 American Chemical Society

oscillators, with a subsequent shift in ω_{SP} depending on how closely they are positioned (Fig. 5.7a).

Coupling is heavily dependent on the direction of polarisation, as mentioned as an influencing factor on ellipsoid polarisations. With the E parallel to the axes of coupling (Fig. 5.7b) the coupling is linear such that ω_{SP} decreases, leading to an increase in λ_{max}. Conversely, excitation perpendicular to the axes of coupling increases ω_{SP} leading to a blue-shift in λ_{max}. In both cases decreasing of the interparticle spacing exaggerates this effect, leading to larger shifts in λ_{max}.

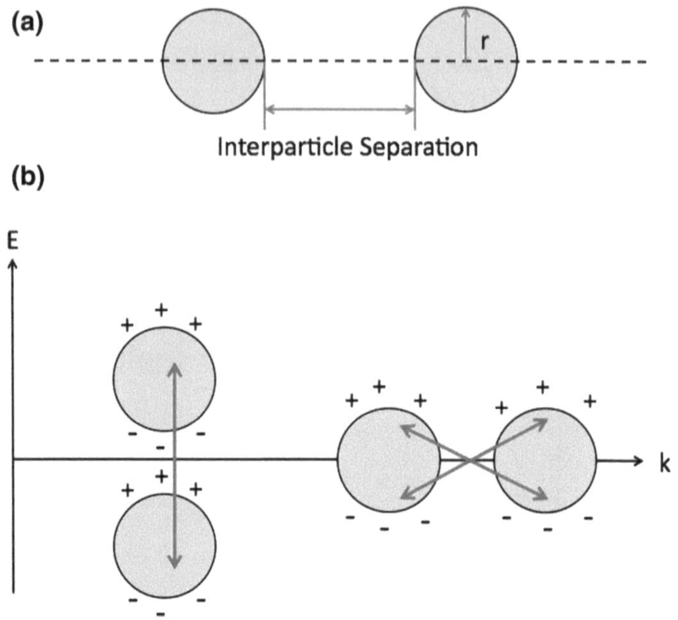

Fig. 5.7 **a** Definition of the interparticle axis and gap for coupled spheres. Electric field polarisation is expressed in relation to this axis. **b** Schematic representation of coupling effects with the interparticle axis aligned parallel (*left*) and perpendicular (*right*) to an electric field; coupled fields are indicated with *blue arrows*

5.2.5 The Discrete Dipole Approximation

In the Discrete Dipole Approximation (DDA) model, a nanoparticle is represented by an assembly of finite cubic elements [44], each small enough that only dipole interactions need be considered. This reduces Maxwell's equations to an algebraic problem for discrete dipoles. The polarisation of each point by a local electric field is given by:

$$P_i = \alpha_i \cdot E_{loc}(r_i) \tag{5.16}$$

where r_i is the location of each cubic element of the divided cluster and E_{loc} is the sum of the contributions from the incident field and all other dipoles:

$$E_{loc}(r_i) = E_{loc,i} = E_{inc,i} + E_{dipole,i} = E_0 e^{[ik.r_i)]} - \sum_{i \neq j} A_{ij} \cdot P_j \tag{5.17}$$

where E_0 is amplitude of the initial wave vector, and $A_{ij} \cdot P_j$ is the interaction matrix, which calculates E_{loc} for all dipoles relative to each other using a radial dependence. It is given by:

$$A_{ij} \cdot P_j = \frac{e^{ikr_{ij}}}{r_{ij}^3} \left\{ k^2 r_{ij}^2 P_j + \frac{1 - ikr_{ij}}{r_{ij}^2} \left[r_{ij}^2 P_j - 3r_{ij}(r_{ij} P_j) \right] \right\} \tag{5.18}$$

where $i \neq j$. For an N dipole system this generates a $3N$-dimensional vector describing the polarisation of all N dipoles in the 3 primary cartesian axes, x, y and z [38]. This is computationally intensive when N is large, thus iterative solutions to the problem have been proposed by Purcell and Pennypacker [44]. Once calculated, P can be used to formulate C_{ext} and C_{abs} in terms of the dipole moments as:

$$C_{ext} = \frac{4\pi k}{|E_0|^2} \sum_{i=1}^{N} (E_{inc,i}^* P_i) \tag{5.19}$$

$$C_{abs} = \frac{4\pi k}{|E_0|^2} \sum_{i=1}^{N} \left\{ Im[P_i(\alpha_i^{-1})^* P_i^*] - \frac{2}{3} k^3 |P_i|^2 \right\} \tag{5.20}$$

where * denotes a complex conjugate. The scattering cross section (C_{sca}) can be obtained from the relationship defined in Eq. 5.8.

5.2.6 DDSCAT

In this work, the SPR spectra have been simulated using the DDA method with the program DDSCAT [11, 45]. DDSCAT 7.0 is an open source Fortran software package which enables the simulation of a variety of nanoclusters via implementation of the DDA, allowing definition of shape, size and composition. For this DDA implementation, the dipole separation d must be small compared to the structural lengths of the incident electromagnetic field. This is satisfied if:

$$|m|kd \ll 1 \tag{5.21}$$

where m is the complex refractive index of the target material ($|m| = n + i\kappa$). Within the volume (V) of the shape being simulated with N polarisable points is, d is:

$$d = \left[\frac{V}{N} \right]^{\frac{1}{3}} \tag{5.22}$$

Moreover, kd quantifies the electromagnetic response of the array of N points, with the change in wave phase corresponding to the lattice spacing. If $kd \to 0$ then $\alpha_i = \alpha_{x,y,z}$ (Eq. 5.12) is valid, however if kd takes a finite value then a correction factor is necessary as outlined by Draine and Goodman [46]: for particles beyond the quasi-static approximation ($r < \lambda/10^4$), increasing particle size shows phase retardation effects. This must be accounted for in α_i, as a finite time is required for an electromagnetic wave to pass induced charge displacement through a particle; accounting for this retardation in the DDA α_i is proposed to be through the lattice dispersion relation [46]:

$$\alpha_i \approx \frac{\alpha^{(0)}}{1 + (\alpha^{(0)}/d^3)[(-4\pi/3)^{1/2}(kd)^2 - (2/3)i(kd)^3]} \tag{5.23}$$

where $\alpha^{(0)}$ is defined as $\alpha_{x,y,z}$ in Eq. 5.12.

A second constraint is that the overall size of the target must be small compared to the wavelength of incident light, ensuring that the quasi-static approximation holds true and only dipolar modes are excited. This is subject to the inequality:

$$r_{eff} < 9.88 \frac{\lambda}{|m|} \left(\frac{N}{10^6} \right)^{\frac{1}{3}} \tag{5.24}$$

where r_{eff} is characterised as the effective radius, calculated as $r_{eff} \equiv (\frac{3V}{4\pi})^{\frac{1}{3}}$. Relevant quantities calculated by DDSCAT are the absorption efficiency factor (Q_{abs}), the scattering efficiency factor (Q_{sca}) and the extinction efficiency factor (Q_{ext}), which are calculated, respectively, from:

$$Q_{abs} \equiv C_{abs}/\pi r_{eff}^2 \tag{5.25}$$

$$Q_{sca} \equiv C_{sca}/\pi r_{eff}^2 \tag{5.26}$$

$$Q_{ext} \equiv Q_{abs} + Q_{sca} \tag{5.27}$$

where C_{abs} and C_{sca} are as previously described the absorption and scattering cross sections for a target, respectively.

5.2.7 Implementation

In our work we have looked at the UV-visible region of the electromagnetic spectrum (300 nm $< \lambda <$ 1,200 nm). Experimental values from thin-film SPR measurements for Au, Pd and Ag in this region are available for the components of the refractive index [47, 48]; for Pt, bulk measurements were used [49]. These values were originally measured linearly in energy (E), and thus the spacing in terms of λ is not consistent, as shown in Fig. 5.8: interpolation across the data range allows more detailed feature definition at high λ (Fig. 5.9b), thus interpolated values are used where indicated in the text.

It is noteworthy here that the refractive indices of the group X elements (Pd, Pt) show no features in the spectrum of interest, whilst the group XI elements Au and Ag show identifiable gradient changes in the imaginary component at ~500 and ~300 nm, respectively, which are features of interband ($d \rightarrow s$) transitions. At these

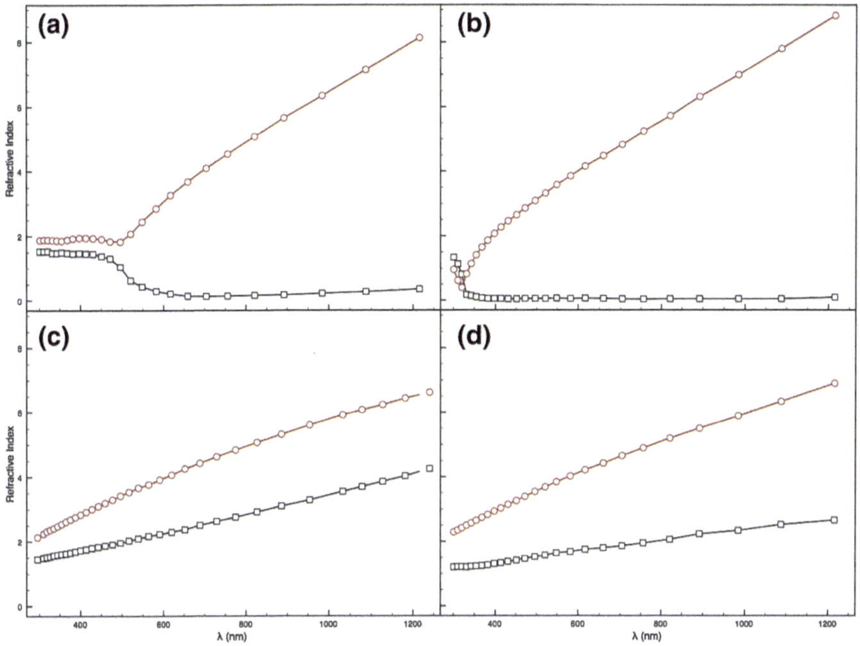

Fig. 5.8 Experimentally measured real (*black*) and imaginary (*red*) components of the complex refractive index for (**a**) Au [47] (**b**) Ag [47] (**c**) Pt [49] and (**d**) Pd [48]. Points are shown from the original data; interpolated values are shown by *fitted lines*. Adapted with permission from Ref. [42]. Copyright 2012 American Chemical Society

same wavelengths the real component of the refractive indices, and thus phase speed of λ in the metal, tends to 0.

Previous work has shown that the DDA is converged to a few percent when the number of dipoles $(N) \geq 10^3$ [25]; values lower than this can cause spurious peaks, as illustrated in Fig. 5.9a. For our calculations dipole spacing was assigned to 1 Å, unless otherwise stated, giving a dipole density of 1,000 dipoles nm^{-3}, and ensuring $N > 10^5$ throughout. This satisfies the criteria given in Eqs. 5.21 and 5.24, for the UV-Visible spectrum.

Calculations were performed on the University of Birmingham BlueBEAR linux cluster, which has 384 dual-core worker nodes, each with two 2.6 GHz AMD processors and 8 Gb memory, and a pool of over 150 Tb of storage space [50].

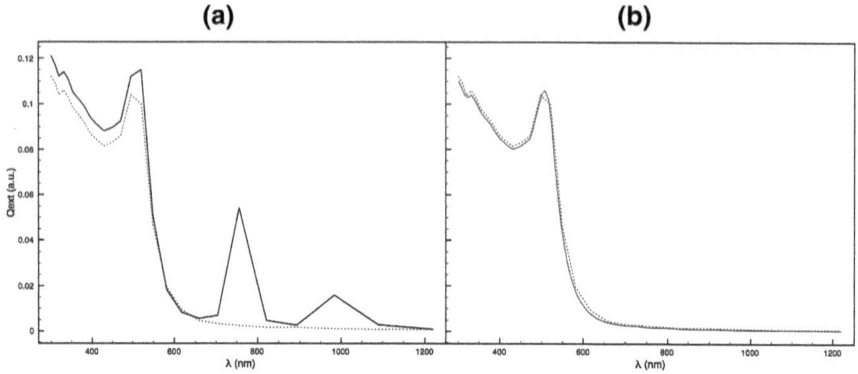

Fig. 5.9 Test calculations simulating the extinction spectra for: **a** An Au nanosphere with radius 2.5 nm, using 81 (*solid, black line*) and 14,328 (*red, dashed line*) dipoles. Spurious peaks are visible for when the dipole resolution is ~1 dipole nm^{-3}, however these are removed at the higher dipole density of ~200 dipoles nm^{-3}, **b** An Au nanosphere with radius 2.5 nm, using literature refractive indices (*red, solid line*) over 27 data points, and interpolated refractive indices (*blue, solid line*) over 120 data points. Reprinted with permission from Ref. [42]. Copyright 2012 American Chemical Society

5.3 Spheres

5.3.1 Monometallic Au and Pd Spheres

Initial calculations of the UV-Visible extinction spectra were performed for spheres of radius r where $2.5\,\text{nm} \leq r \leq 10\,\text{nm}$, incrementing in 2.5 nm steps, for Au and Pd; the results are plotted in Fig. 5.10. Pd appears featureless in the UV-visible region with decreasing Q_{ext} as λ increases; however Au has a peak close to 510 nm associated with the interband transitions observed in the refractive index data (Fig. 5.8); this is consistent with previous reports [15, 16].

As r increases up to 10 nm, so the resonance intensity for both Au and Pd also increases; λ_{max} remains relatively constant, though the accuracy with which the position can be identified is limited by the selected refractive indices used. A non-linear red-shift would be expected for λ_{max} with increased $r > 10\,\text{nm}$ [23, 37], however the lack of accuracy from the limited experimental measurements of the refractive index makes it difficult to identify shifts in λ_{max} at the sizes we are investigating. We can however deduce that any shift in λ_{max} is small at this size. The difference between the use of the original 27 experimental data points and the interpolated refractive indices is given in Fig. 5.9b for a sphere with $r = 2.5\,\text{nm}$, and clearly the peak of $\lambda_{max} = 508\,\text{nm}$ is more perceptible.

Also of note in Fig. 5.10a is the decreasing width of extinction features as r increases. This is quantitatively referred to as the "full width, half maximum" (FWHM) and reflects the dephasing time of a particle once it has been influenced

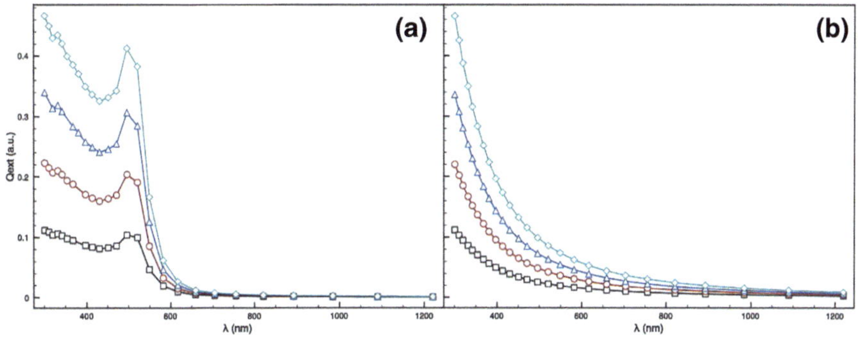

Fig. 5.10 SPR spectra calculated for (**a**) Au and (**b**) Pd spheres. Radii of 2.5, 5, 7.5 and 10 nm are shown in *black*, *red*, *blue* and *green*, respectively. Refractive indices have been taken directly from the literature [47, 48]

by a photon. For small particles ($r < 20$ nm), the FWHM is influenced by electron damping from surface-scattering, and further intrinsic size effects, whilst in larger particles ($r > 20$ nm) radiative damping creates broadening. For clarity, the extrinsic size effect is the excitation of higher order modes (e.g. quadrupolar) in larger particles, and influences the position of λ_{max} [40].

In Fig. 5.11 the contributions to Q_{ext} from both Q_{sca} and Q_{abs} at λ_{max}, as a function of r, are given. Absorption dominates at small sizes: in our plot Q_{abs} is greater than Q_{sca} for $r = 2.5$ nm. However, as r increases, Q_{abs} increases pseudo-linearly whilst Q_{sca} grows more like an exponential function, such that for $r =$ 10 nm: $Q_{sca} > Q_{abs}$. This trend is known to continue on to larger sizes, where scattering effects dominate extinction [40].

5.3.1.1 Coupled Au Spheres

Figure 5.12 shows the extinction spectra for two closely coupled Au nanospheres ($r = 2.5$ nm), with polarisation calculated perpendicular and parallel to the interparticle axis. Interparticle separation was varied systematically between 0 (touching) and 2.5 nm (i.e. a radius distance away). Contrasting spectral effects are seen to be dependent on the orientation of the interparticle axis relative to the direction of k.

With k parallel to the interparticle axis (Fig. 5.12a), decreasing interparticle separation is found to decrease the intensity of Q_{ext} in a quasi-linear manner, with λ_{max} being seemingly blue-shifted from 521 nm towards 496 nm; at interparticle separation of 0 nm the Q_{ext} spectra is noticeably shifted compared to an isolated Au particle.

In contrast, when k is aligned orthogonal to the interparticle axis (Fig. 5.12b) increased intensity is calculated for Q_{ext} with decreasing interparticle separation. For an interparticle separation of 2.5 nm, the extinction spectrum is akin to a superposition of two Au particles, however at distances <1 nm a distinct shift in λ_{max} is seen, with

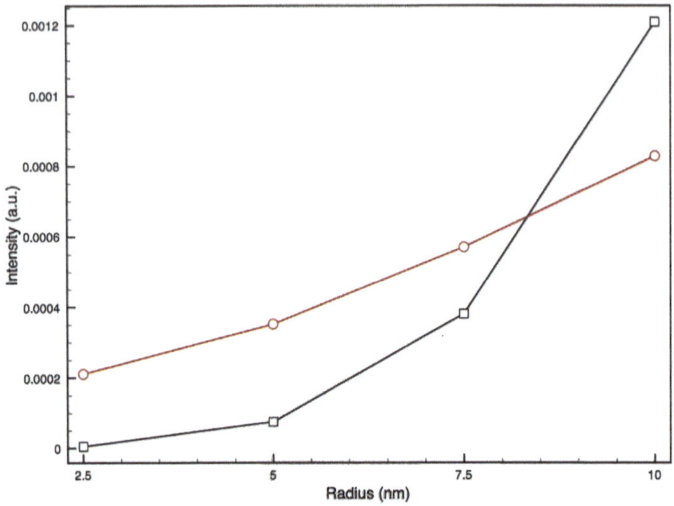

Fig. 5.11 Contributions to Q_{ext} from Q_{sca} (*black squares*) and Q_{abs} (*red circles*), for an Au sphere with increasing sphere radius

non-linear growth in intensity. The strong coupling for an interparticle separation of 0 nm is similar to the LSPR shifts in λ_{max} found in nanorods, with a dependence on the AR. When the interparticle separation is $>5r$, standard Mie theory should apply [23].

The coupling mechanisms are due to induced dipolar interactions in the two nanospheres, as documented in Sect. 5.2.4. The non-linearity found at small inter-particle distances is hypothesised as a result of sudden charge redistribution within the particle volume once beyond a coupling threshold, to ensure intraparticle neutrality, thus effecting the polarisability. In a similar manner, the non-linear red-shift in λ_{max} for nanospheres coupled perpendicular to k appears quasi-exponential at small interparticle gaps due to the instantaneous charge equilibration. Dependence on the ratio of gap size to r is postulated, however clear distinctions are still not available in the literature; consideration of higher excitation modes may also be a possibility, though the DDA is not suitable for investigating this further [23].

5.3.2 *Bimetallic Core-Shell Spheres*

5.3.2.1 Au-Pd

Results are presented in Fig. 5.13 for bimetallic $Au_{core}Pd_{shell}$ and $Pd_{core}Au_{shell}$ particles. Using a seed particle as the core, with radius (r_{core}) of 2 nm, and refractive index ε_{core}, a secondary metal, with refractive index ε_{shell}, was introduced as a

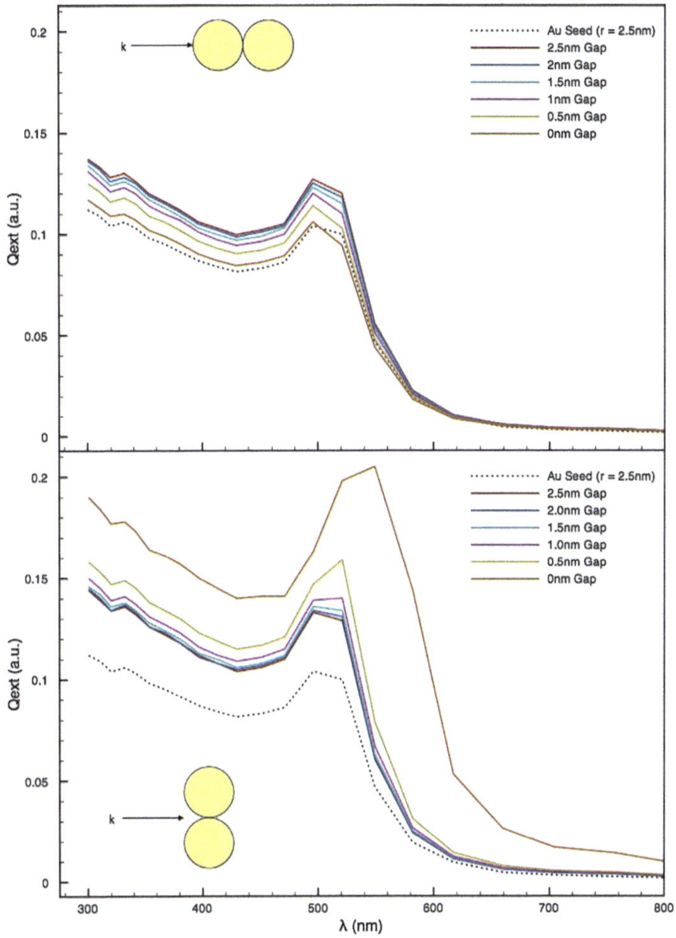

Fig. 5.12 Q_{ext} for coupled Au nanospheres with a radius of 2.5 nm each. *Top* SPR excitation with k parallel to the interparticle axis. *Bottom* SPR excitation with k perpendicular to the interparticle axis. An Au seed particle, with radius of 2.5 nm, is plotted with a *black*, *dashed line*. Interparticle separations of 2.5, 2, 1.5, 1, 0.5 and 0 nm are plotted in *red*, *blue*, *green*, *purple*, *olive* and *brown*, respectively, as shown in the legend. Refractive indices have been taken directly from the literature [47, 48]

surface covering with increasing shell thickness (r_{shell}) from 0.1 to 2 nm, giving a maximum combined radius of 4 nm. A 0.1 nm shell corresponds to a monolayer of shell dipoles in our calculation, however the experimental radius of Au and Pd atoms is slightly greater than this, at 1.35 and 1.4 Å [51], respectively, and must be considered when analysing results.

For an Au core, we have already shown a strong extinction feature exists at $\approx$510 nm in Sect. 5.3.1. The addition of a thin layer of Pd ($r_{shell} = 0.1$ nm) results in

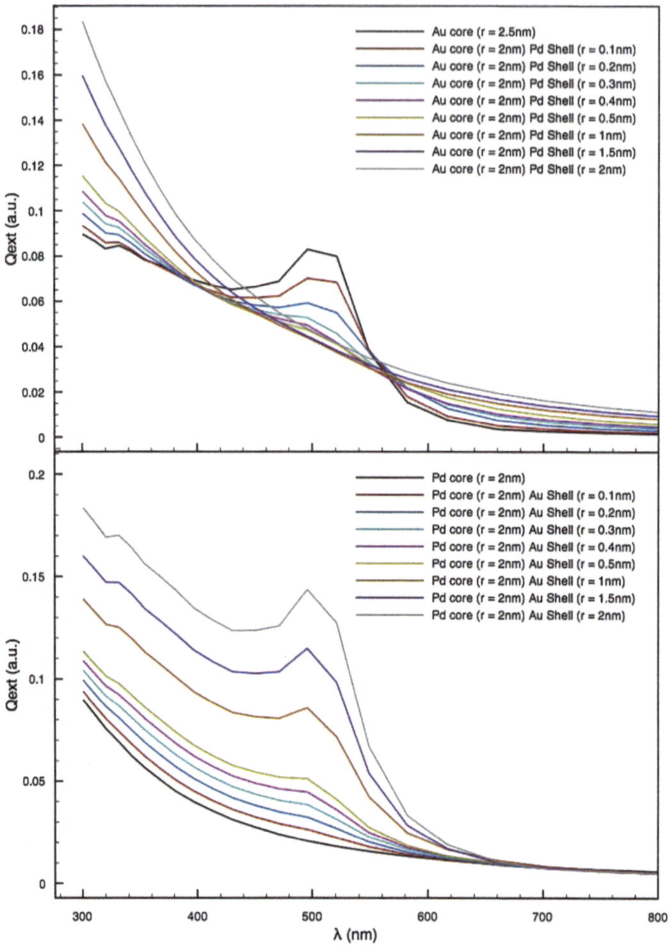

Fig. 5.13 SPR spectra calculated for $Au_{core}Pd_{shell}$ (*top*) and $Pd_{core}Au_{core}$ (*bottom*) spheres. In both cases an initial core of 2 nm is used and then shells of incrementing thickness were added (up to 2 nm). The absorption spectrum for an Au seed particle with $r = 2$ nm is provided for reference with a *black*, *solid line*. A legend is provided. Refractive indices have been taken directly from literature

damping of this feature, and increase in the FWHM; there is no identifiable shift in λ_{max} from the raw data, yet interpolation of the Q_{ext} results yields a slight blue-shift. Continued addition of dipole monolayers up to 0.5 nm Pd shell thickness results in stronger quenching of the Au spectral feature at $\approx$510 nm, with the original Au feature barely identifiable with $r_{shell} = 0.5$ nm. The change in Q_{ext} is rapid (non-linear) up to a shell thickness of 0.5 nm; thicker Pd shells ($r_{shell} > 1$ nm) fully dampen the Au feature.

By contrast, for a Pd core with increasing Au shell thickness, the appearance of the Au spectral feature at ≈ 510 nm occurs at a quasi-linear rate. For a shell monolayer ($r_{shell} = 0.1$ nm), a slight shoulder appears on the Q_{ext} spectrum. This improves in a sequential manner up to a shell thickness of 0.5 nm. where the trademark Au feature is discernible. For thicker shell sizes ($r_{shell} > 1$ nm) λ_{max} does not shift, however the intensity increases and the FWHM decreases with additional shell layers, similar to observations for monometallic Au spheres (Fig. 5.10a). The prominence of the Au feature is clear when $r_{shell} = 2$ nm; a thicker Au shell is required to produce this feature on a Pd core than is required for a Pd shell to dampen the Au features.

5.3.3 *Off-Centred Bimetallic Core-Shell Spheres*

5.3.3.1 Au-Pd

Particles with an off-centred core are formed when shell growth is not uniform across a seed particle. For a core-shell sphere with a fixed core of $r = 2$ nm, the volume ratio of core to shell for stoichiometries ≈ 1:1 and 1:4 is equal to $r_{shell} =$ 0.5 nm and $r_{shell} = 1.5$ nm, respectively. Taking these size parameters as standard, calculations were performed to measure the effect core displacement, within the core-shell nanoparticle, has on the overall extinction spectra: results are plotted in Fig. 5.15 for both $Au_{core}Pd_{shell}$ and $Pd_{core}Au_{shell}$. To ensure clarity in potentially identifying TSPR and LSPR features, the incoming angle, θ, of the incident radiation k is set to 45° relative to the axis of core displacement (Fig. 5.14).

The Q_{ext} spectra for both $Au_{core}Pd_{shell}$ and $Pd_{core}Au_{shell}$ spherical particles are plotted in Fig. 5.15, with $r = 2.5$ and 3.5 nm. Resulting extinction spectra are shown for the core centred in the particle and with the core offset along the x-axis such that it was touching the particle surface. Calculations were performed with incremental core-displacements in 1 Å steps (i.e. 1 dipole), however we have only plotted the two extremes. In general we found that only when $r_{shell} < 0.5$ nm (at the thinnest shell covering) were any changes in the spectra identifiable; for $r_{shell} \geq 0.5$ nm the Pd forms too thick a layer.

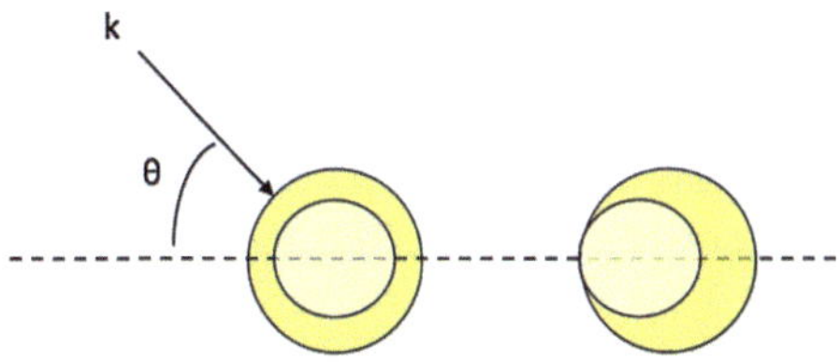

Fig. 5.14 Schematic to represent the setup for off-centred core-shell particles, with the core (*pink*) and the shell (*yellow*) displayed. The *left-hand particle* is core-centred, whilst the *right-hand particle* has the core displaced along the x-axis. The angle of incident radiation, θ, is also illustrated

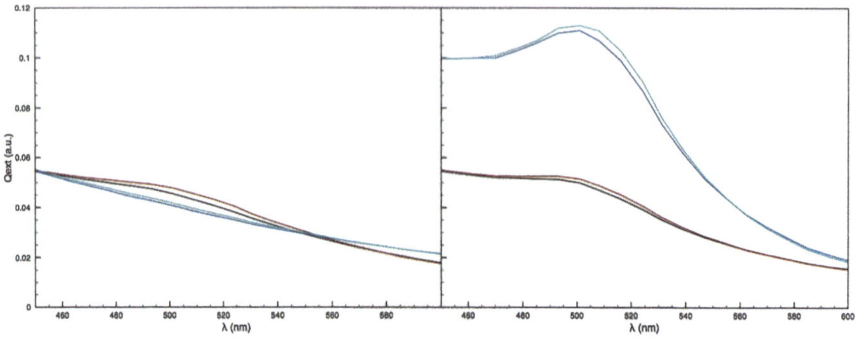

Fig. 5.15 *Left* Q_{ext} for $Au_{core}Pd_{shell}$ nanoparticles. *Right* Plot of Q_{ext} for $Pd_{core}Au_{shell}$ particle. For particles with an overall radius of 2.5 nm, Q_{ext} is plotted for a core-centred particle and core-displaced particle in *black* and *red*, respectively. For particles with an overall radius of 3.5 nm, Q_{ext} is plotted in *blue* and *green* for core-centred and core-displaced particles, respectively. Refractive indices have been interpolated from the literature. Reprinted with permission from Ref. [42]. Copyright 2012 American Chemical Society

For $Au_{core}Pd_{shell}$, we have already seen that much of the trademark Au feature at $\approx$510 nm has been quenched by a Pd shell of 0.5 nm thickness. However, with $r_{shell} = 0.5$ nm, we see that shifting the Au core towards the edge of the particle gives rise to an increase in intensity relative to the core-centred particle; though the intensity increase on the gentle peak is only slight. Increasing r_{shell} to 1.5 nm, we find that there is also an increase in intensity around λ_{max} with core displacement, though the relative increase compared to the core-centred particle is half that seen for core displacement when $r_{shell} = 0.5$ nm.

Likewise, for $Pd_{core}Au_{shell}$, displacement of the Pd core towards the particle edge slightly increases the intensity on the Au spectral feature. For $r_{shell} = 0.5$ nm, the intensity increase is similar to that seen for $Au_{core}Pd_{shell}$, with the spectra of the two inverse heterogeneous arrangements being very similar: if anything the extinction peak is marginally more pronounced for $Pd_{core}Au_{shell}$. When r_{shell} is increased to 1.5 nm, not only does the extinction intensity at λ_{max} increase with core displacement, but also the intensity shift is more pronounced; the FWHM also decreases slightly. When the Pd core is shifted to the particle edge then the Au volume has a more direct pathway, through ε_{Au}, for plasmon resonances to form.

Investigation into the influence of θ, with results plotted in Appendix D.1, show there is minimal dependence on θ for the resonance intensity, and no dependence on the location of λ_{max}. Intensity values are at an equivalent minimum when $\theta = 0°$ or 180°, and at a maximum when $\theta = 90°$. This means that though our observation of intensity increasing at λ_{max} is valid for core displacement as a greater direct volume of Au is available as a plasmon pathway, extinction is intensified if this Au concentration is on one distinct side of the particle relative to the oscillating electric field.

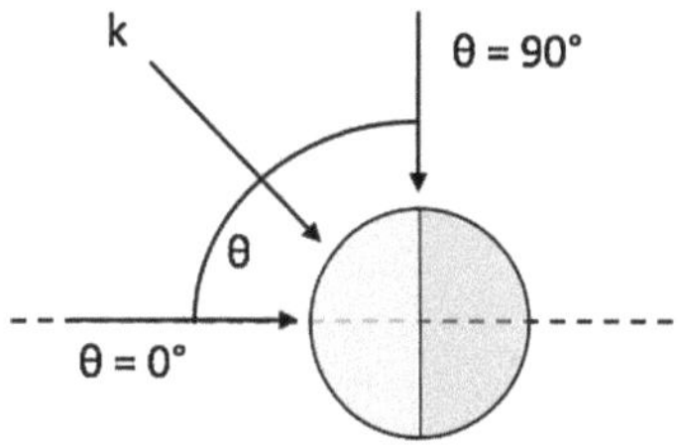

Fig. 5.16 Depiction of a Janus particle, with various orientations of the incoming angle (θ) for the electromagnetic radiation (k) relative to the normal of the interface. Adapted with permission from Ref. [42]. Copyright 2012 American Chemical Society

Janus particles are formed when bimetallic nanoparticles form with distinct atomic separation, giving a surface with two (or more) sets of properties: resulting in two halves with different dielectric values [52]. By definition, with a perfect split they have 1:1 volume stoichiometries (Fig. 5.16). Taking a particle with $r = 2.5$ nm, we have performed calculations for the extinction spectrum of such a particle, as plotted in Fig. 5.17. We find that the Q_{ext} is greater in intensity for Janus particles than core-shell segregated particles; both for $Au_{core}Pd_{shell}$ and $Pd_{core}Au_{shell}$. We also find that for Janus particles θ has no influence on the intensity of Q_{ext}, nor λ_{max}. An assumption can therefore be made that a maximised mean free pathway through the Au, such as when $\theta = 0°$, is equivalent to the affect of increased Au concentration on one side of the particle relative to the exterior electric field, such as when $\theta = 90°$; this is not the case for the off-centred particles, in our calculations, as the mean free path through Au is never equal to the diameter of the particle.

5.3.3.2 Au-Ag

Previously, experiments conducted by Li et al. [36] and accompanying calculations performed by Phillips [53] have proven informative with regards to associating the effect of increasing shell thickness on core-centred $Au_{core}Ag_{shell}$ particles, and the inverse $Ag_{core}Au_{shell}$ structure, with experimental observations: the DDA calculations showed a strong relationship between shell thickness and the position of λ_{max}, with the strength of the Ag extinction features, located near 350 nm, greater than the Au equivalent defining features (≈510 nm). For pure Au and Ag spherical particles, with $r = 2.5$ nm, Q_{ext} is ten times greater in the latter [38, 53]. In our work calculations were performed using our off-centred arrangements to see if further insight could be achieved on this optically challenging area.

Calculations for $Au_{core}Ag_{shell}$ and $Ag_{core}Au_{shell}$ spheres, with a fixed overall radius of 2.5 nm, have been used to produce the plots of Q_{ext} provided in Fig. 5.18 with varying degrees of core displacement along the x-axis. In both cases, the Ag spectral feature dominates, however a shoulder associated with Au is present at ~500 nm. Previous work has shown that an Au shell has a damping effect on the Ag

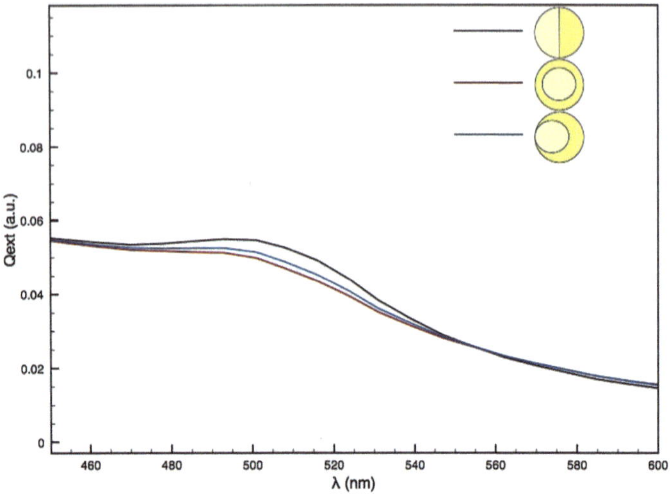

Fig. 5.17 Plot of the extinction coefficient, Q_{ext}, for a Janus particle (*black*) compared to a $Pd_{core}Au_{shell}$ core-centred particle (*red*) and a core-displaced particle (*blue*). A legend is provided, with Pd and Au represented in *white* and *yellow*, respectively. An incident radiation angle of $\theta = 45°$ was used. Refractive indices have been interpolated from literature values. Adapted with permission from Ref. [42]. Copyright 2012 American Chemical Society

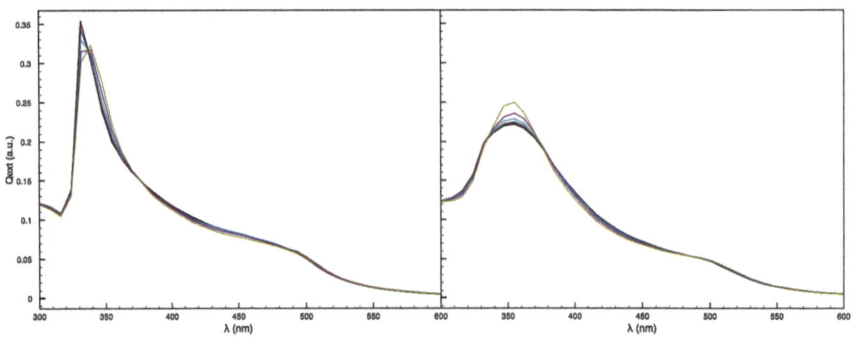

Fig. 5.18 *Left* Q_{ext} for $Au_{core}Ag_{shell}$ spheres with increasing displacement of the core. *Right* Q_{ext} for $Ag_{core}Au_{shell}$ with core displacement. The core-centred particle is plotted in *black*, with displacements of 0.1, 0.2, 0.3, 0.4 and 0.5 nm (i.e. up to touching the surface) plotted in *red*, *blue*, *green*, *purple* and *olive*, respectively. Refractive indices have been interpolated from the literature. Adapted with permission from Ref. [42]. Copyright 2012 American Chemical Society

core, up to $r_{shell} \approx r_{core}$, whilst for an Ag shell on an Au core a monolayer shell of Ag ($\approx$0.2 nm) quenches Au spectral features and produces extinction features distinguishable as those of Ag [28, 53].

Closer attention to the position and intensity of the Ag spectral feature, λ_{max}, as plotted in Appendix D.2, finds that both are dependent on the core-displacement. Q_{ext} at λ_{max} is strong, in agreement with previous work [40]. For $Au_{core}Ag_{shell}$, the value of Q_{ext} at λ_{max} decreases linearly with core displacement between 0.1 and 0.5 nm; when the shell thickness is less than 0.1 nm (i.e. displacement of 0.4 nm) λ_{max} shifts to 340 nm. Whilst the intensity at λ_{max} appears to decrease, this may just be an artefact of the shift in position to a wavelength unsampled in calculation; interpolation of the results offers some insight into the shift of λ_{max} being relatively smooth.

Core-displacement in the $Ag_{core}Au_{shell}$ sphere does not result in the same perceptible changes in λ_{max}; however the extinction spectra is not as sharp around λ_{max} and it is likely this makes identifying changes more difficult. The intensity of Q_{ext} increases with the core-displacement; this no doubt is due to the stronger plasmon coupling with the Ag dielectric medium as the core approaches the surface of the nanoparticle.

5.3.4 Conjoined Bimetallic Core-Shell Spheres

5.3.4.1 Au-Pd

Experimental work by Cookson [54] found an unusual phenomenon whilst creating $Au_{core}Pd_{shell}$ particles, whereby a red-shift in λ_{max} was observed, contrary to calculations for isolated $Au_{core}Pd_{shell}$ spheres as described in Sect. 5.3.3.1. It was suspected that polydispersity, partial aggregation or irregular-shaped particles were influencing the particle formation: we give an overview of the observations by Cookson [54], before detailing DDA calculations performed in an attempt to offer some understanding to this phenomena.

5.3.4.2 Experimental

UV-Visible absorption spectra for isolated Au seeds (Fig. 5.19) show strong absorption at 520 nm. This feature is characteristic of Au spheres on the nanometre scale and has been well documented previously [16, 23]. Addition of a Pd shell results in a red shift of the main feature (λ_{max}) to 550 nm.

HAADF-STEM imaging shows that $Au_{core}Pd_{shell}$ particles have formed. Due to Z-contrast the brighter Au core ($Z = 79$) is distinguished from the darker Pd shell ($Z = 46$) (Fig. 5.20c); however merging of these Pd shells is visible, corresponding to conjoined $Au_{core}Pd_{shell}$ particles. Further HAADF-STEM analysis shows that the $Au_{core}Pd_{shell}$ nanospheres have formed chains, many particles in length, and that chain dispersion increases with heating time up to 8 h (Fig. 5.20a, b).

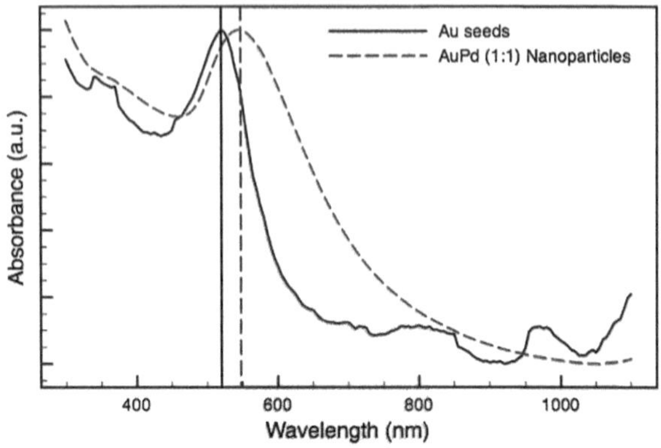

Fig. 5.19 Experimental UV-Visible absorption spectra for Au seeds and $Au_{core}Pd_{shell}$ nanoparticles formed after 8 h heating, normalised to λ_{max} (indicated by *vertical lines*). Reprinted with permission from Ref. [55]. Copyright 2010 American Chemical Society

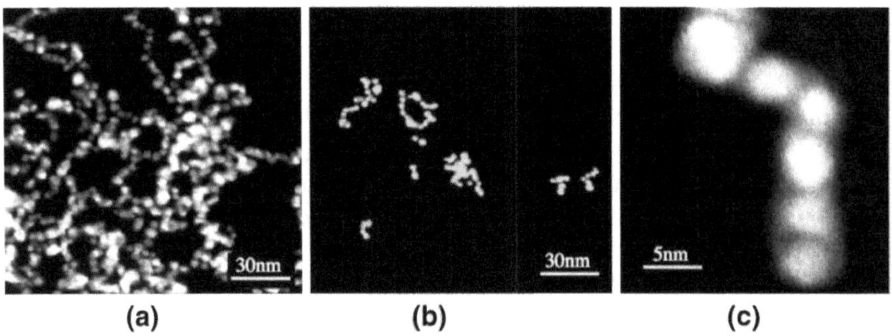

Fig. 5.20 *Left* to *right* HAADF-STEM images of $Au_{core}Pd_{shell}$ particles after (**a**) 30 min and (**b**) 8 h heating time, respectively. (**c**) A higher resolution image of part of a $Au_{core}Pd_{shell}$ particle chain formed after 8 h heating, illustrating the $Au_{core}Pd_{shell}$ segregation. Reprinted with permission from Ref. [55]. Copyright 2010 American Chemical Society

5.3.4.3 Simulation

DDSCAT allows user-defined arbitrary shapes, which we utilise in this study to replicate the merging of spherical $Au_{core}Pd_{shell}$ nanoparticles witnessed experimentally. Geometries were created representing two and three collinear spherical particles with touching shells, overlapping shells and touching cores (Fig. 5.21).

A radius of 2 nm is used for the Au core of the spherical particles throughout, with a Pd shell of thickness 0.5 nm, giving an overall radius of 2.5 nm (as measured from the STEM images). For isolated spheres this corresponds to a volume stoichiometry $\approx$1:1. For a single $Au_{core}Pd_{shell}$ particle the Au SPR features are quenched considerably by the addition of a Pd coating (Fig. 5.22) [31], however a small ripple remains at $\approx$500 nm. Initially, calculations were carried out *in vacuo* ($\varepsilon_m = 1$) for

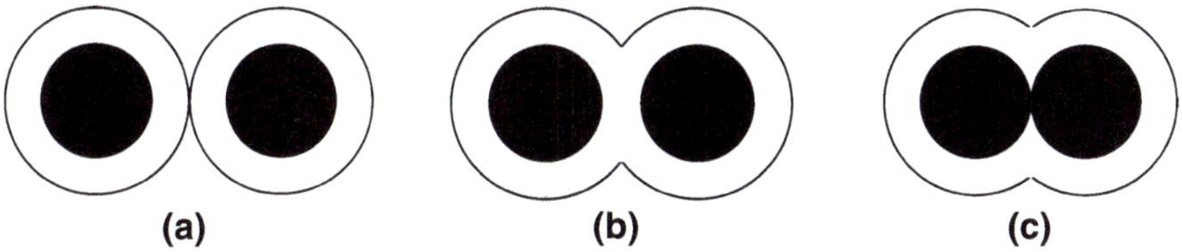

Fig. 5.21 *Left* to *right* Schematics of the models used for (**a**) touching shells, (**b**) shell-overlap and (**c**) touching cores. *Black* and *white* shading represent the Au core and Pd shell, respectively. Reprinted with permission from Ref. [55]. Copyright 2010 American Chemical Society

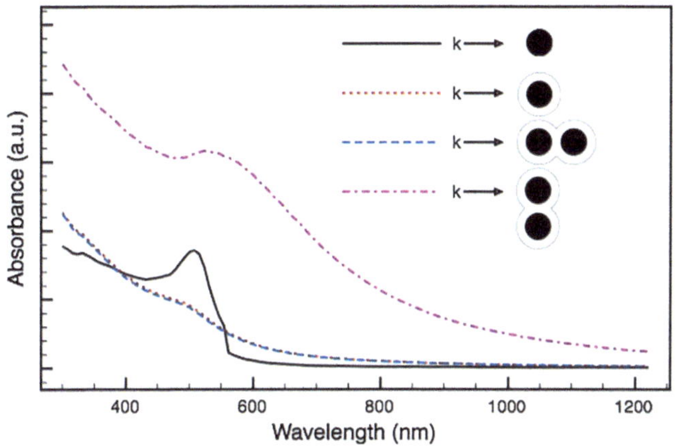

Fig. 5.22 DDSCAT calculations for isolated Au and $Au_{core}Pd_{shell}$ spheres compared to $Au_{core}Pd_{shell}$ particles oriented with conjoining axis parallel and perpendicular to k. Au and Pd are represented in the schematics by *black* and *white*, respectively. Refractive indices have been interpolated from the literature. Reprinted with permission from Ref. [55]. Copyright 2010 American Chemical Society

two shell-overlapping spheres with the interparticle axis parallel (0°) and perpendicular (90°) to the incident radiation (k) to identify the potential SPR contributions of longitudinal (LSPR) and transverse (TSPR) waves (Fig. 5.22).

The prominent contribution to the UV-Visible absorption spectra is shown to be from the LSPR, but TSPR influences cannot be discounted in our calculations. Previously, red shifts in LSPRs have been witnessed with increasing AR of Au rods [13, 17–19] and also coupling of spheres [16]. For the case of conjoined particles the length is defined along the interparticle axis whilst the width is equal to the diameter of a single particle.

In our calculations the TSPR feature neither moves nor intensifies, implying the change is not associated with coupling effects. For coupled particles (i.e. not touching) overall charge neutrality must be respected for each particle; in our case the overlapping dielectric Pd shells mean the intraparticle charge neutrality does not have to be respected for the Au cores, thus the particle overlap results in nanorod-like features (dependent on the AR) with longitudinal plasmon oscillations [56].

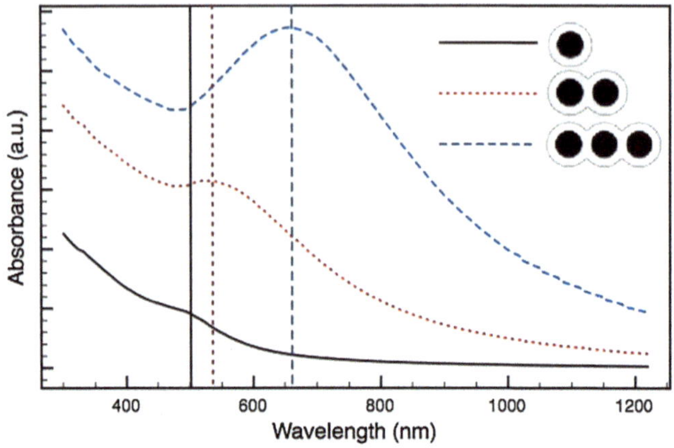

Fig. 5.23 DDA calculated SPR spectra for increasing numbers of conjoined $Au_{core}Pd_{shell}$ particles. λ_{max} is indicated by vertical lines of matching key. Au and Pd are represented in the schematics by *black* and *white*, respectively. Refractive indices have been interpolated from the literature. Reprinted with permission from Ref. [55]. Copyright 2010 American Chemical Society

Henceforth, the radiation source is set at $\theta = 45°$ to the interparticle axis to account for contributions from both the LSPR and TSPR.

Increasing the number of conjoined particles in the chain results, intuitively, in increased ARs: for one, two and three conjoined particles, with overlapping shells, the ARs are 1, 1.9 and 2.8 respectively. In turn this increase results in a red shift of λ_{max}. However, the increase in λ_{max} is non-linear with AR: between two and three particles the red shift is 120 nm, considerably greater than the red shift of 40 nm between one and two particles (Fig. 5.23). This non-linearity is due to the SPR dampening effect of the Pd shell between the Au cores, as previous reports for Au nanorods give linear relationships [17–19] up to a width of 30 nm, at which point the quasi-static approximation no longer remains valid [57].

Varying the degree of overlap between two $Au_{core}Pd_{shell}$ particles, using the models previously described, we see changes in λ_{max} (Fig. 5.24a). Initially, for touching particles, the extinction spectrum is a superposition of the spectra for individual nanospheres; whilst the intensity has increased, the resonance shape and position is identical to an isolated $Au_{core}Pd_{shell}$ particle ($\lambda_{max} = 500$ nm). This can be explained by the quenching effect of Pd dominating the spectrum, as it shields the Au cores from each other. Overlapping of the particles causes a red shift of λ_{max} to 528 nm. The FWHM is large due to Pd shielding of the Au cores. Further overlap, to the extent that the Au cores are touching, results in a blue shift of λ_{max} back to 516 nm due to the decreased AR. Here the shift is accompanied by a decrease in the FWHM. This increase in definition and intensity is attributable to the reduced levels of Pd, and thus shielding, between the Au cores.

Calculations for three conjoined $Au_{core}Pd_{shell}$ particles yield similar results (Fig. 5.24b). Touching particles again represent a superposition, with $\lambda_{max} \approx 500$ nm.

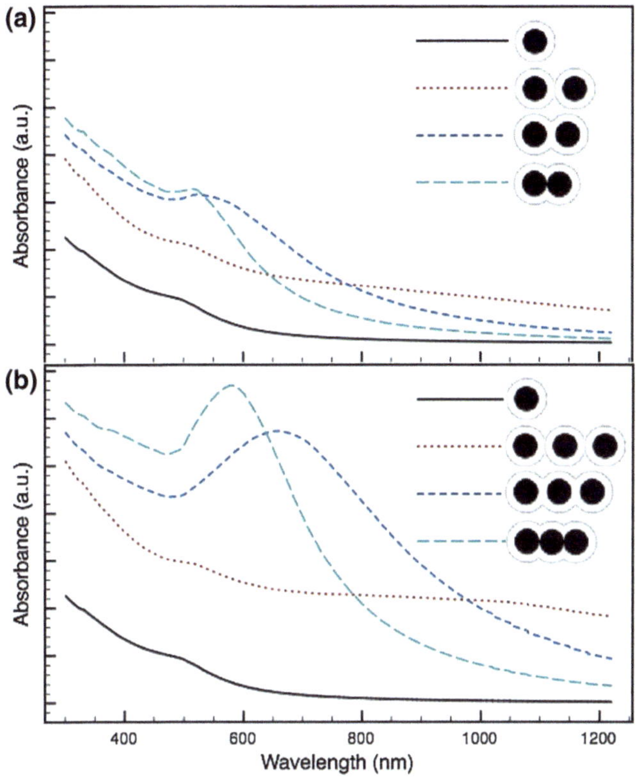

Fig. 5.24 (**a**) Calculated SPR extinction spectra for differing levels of overlap between two conjoined $Au_{core}Pd_{shell}$ particles compared to (**b**) similar calculations for three conjoined $Au_{core}Pd_{shell}$ particles. Refractive indices have been interpolated from the literature. Reprinted with permission from Ref. [55]. Copyright 2010 American Chemical Society

Shell overlapping results in a dramatic red shift of λ_{max} to 658 nm, accompanied by peak broadening. Bringing the Au cores into contact with each other leads to λ_{max} blue shifting back to 578 nm, with increased intensity and decreased FWHM. The greater spread displayed by λ_{max} for three conjoined particles can be explained by the increased chain length, and thus AR, leading to longer wavelength LSPR oscillations. Overall, for three particles, intensity is significantly increased across the spectrum compared with two particles.

Calculations were repeated defining the surrounding medium as water ($\varepsilon_m = 1.333$) and toluene ($\varepsilon_m = 1.497$). Previously, the increasing dielectric of a surrounding medium has been found to cause red shifting of the spectra relative to vacuum [38], and our calculations are consistent with this (Fig. 5.25).

For two shell-overlapping $Au_{core}Pd_{shell}$ particles we see a significant red-shift, broadening and quenching of the LSPR feature (Fig. 5.25a). However, in all calculations a peak remains at $\approx$510 nm. This can be attributed to the increasing strength of

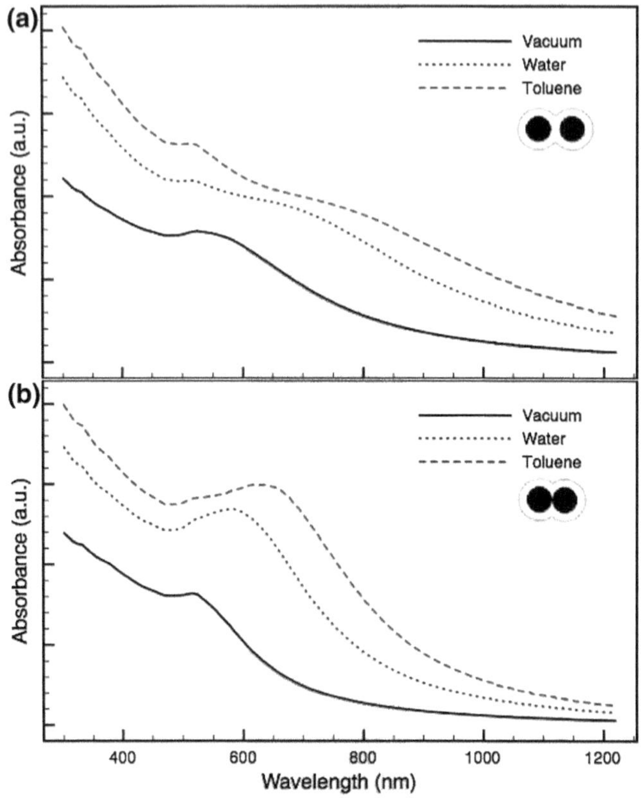

Fig. 5.25 DDA Calculated SPR extinction spectra for two $Au_{core}Pd_{shell}$ particles with (**a**) shell-overlapping and (**b**) core-touching, with various surrounding medium dielectrics (*vacuum*: $\varepsilon_0 = 1$, *water*: $\varepsilon_0 = 1.333$, *toluene*: $\varepsilon_0 = 1.497$). Refractive indices have been interpolated from the literature. Reprinted with permission from Ref. [55]. Copyright 2010 American Chemical Society

the TSPR; calculations for Au cores on their own show a quasilinear growth in intensity with ε_m values. This is contrary to our previous observations, where the TSPR contributed little to the overall spectrum (Fig. 5.22), but implies that the increasing ε_m value results in stronger dipolar interactions with the Pd shell, consequently decreasing its damping effect on the Au core.

For two core-touching spheres there is a well defined peak, with λ_{max} of 516, 578 and 624 nm for vacuum, water and toluene, respectively. Similarly, for three core-touching spheres λ_{max} is 578, 716 and 793 nm for vacuum, water and toluene, respectively. These shifts show a close to linear variation with increasing ε_m.

The effect of the Pd shell on the SPR spectra was tested with models of core radius 2 nm and Pd shell of thickness 1.5 nm; hence a combined radius of 3.5 nm and volume stoichiometry ≈1:4 (Fig. 5.26). Here we found that the overweighting of Pd resulted in complete quenching of the Au features seen previously for thinner shell thickness,

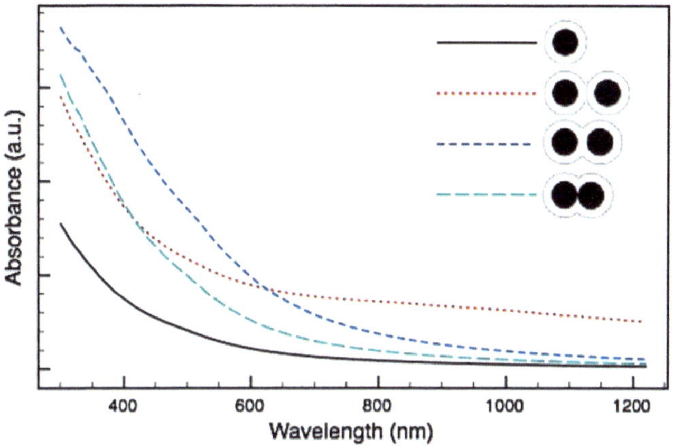

Fig. 5.26 DDA calculated SPR extinction spectra for conjoined $Au_{core}Pd_{shell}$ particles with Au:Pd volume stoichiometry ≈1:4. Au and Pd are represented in the schematics by *black* and *white*, respectively. Refractive indices have been interpolated from the literature. Reprinted with permission from Ref. [55]. Copyright 2010 American Chemical Society

and individual $Au_{core}Pd_{shell}$ particles (Figs. 5.23 and 5.24). We conclude that the Pd shell, therefore, is thick enough for a volume stoichiometry of 1:4 that only Pd SPR features are prominently visible. This effect has been observed, on occasions, experimentally in the work of Cookson [54], and is consistent with previous studies of isolated $Au_{core}Pd_{shell}$ nanospheres where a Pd coating was found to quench features in the UV-Visible region [31].

Figure 5.27 plots λ_{max} vs. AR as previously documented for Au rods [17, 58] compared with our results for $Au_{core}Pd_{shell}$ particles with 1:1 volume stoichiometry. For touching $Au_{core}Pd_{shell}$ particles we have an almost flat line, with minimal coupling-based interactions. A completely horizontal line would be expected for non-interacting particles as there is no distinguishability between the TSPR and LSPR. Similarly there is only a small λ_{max} red shift for two core-touching and shell-overlapped particles (Fig. 5.24a). Only for three conjoined $Au_{core}Pd_{shell}$ spheres do we see a significant increase of λ_{max} (Fig. 5.24b), which we can attribute to the LSPR. The red shift of λ_{max} is consistent in part with our preceding work, and previous work by Zhong et al. on collinear Au spheres who found that separation of the TSPRs and LSPRs, and a red shift of the main SPR, were visible when Au spheres were brought together in the simulation [59].

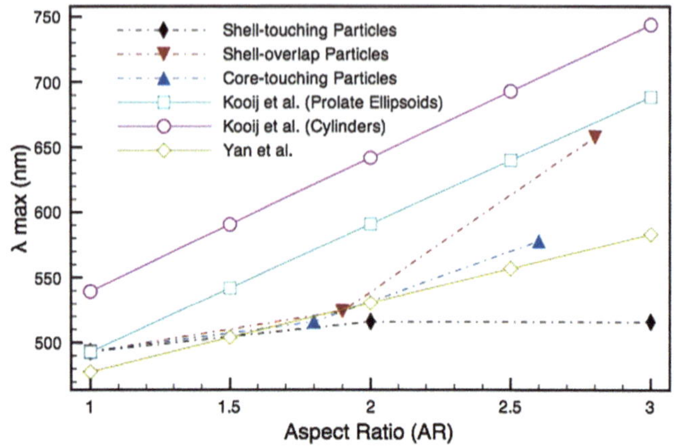

Fig. 5.27 Comparison of the position of λ_{max} in our results for $Au_{core}Pd_{shell}$ particles with 1:1 volume stoichiometry with previous work on Au nanorods by Kooij et al. on prolate ellipsoids and cylinders [13], and by Yan et al. on hemispherically capped cylinders [58]. Reprinted with permission from reference [55]. Copyright 2010 American Chemical Society

5.4 Ellipsoids

5.4.1 Monometallic Au Ellipsoids

The position of λ_{max} can be manipulated by variation of the particle AR, where one axis is elongated relative to the other two, which remain equivalent. Distortion of a sphere in this manner forms a prolate ellipsoid. Calculations were performed for Au prolate ellipsoids to investigate the effect of AR on Q_{ext}, with normalised results plotted in Fig. 5.28. For the short axis $b = c = 5$ nm, whilst the x-axis, with length a, was varied between 5 and 30 nm (ηr, where η = AR); the angle of incident radiation was set to $\theta = 45°$ relative to the long-axis.

A strong red-shift of λ_{max} is seen with increasing AR, in agreement with previous work on ellipsoids [13] and other rod shapes [18]. This feature is attributed to the LSPR of the rod-shaped particle; a residual feature at ~510 nm of weaker intensity is a result of the lower intensity TSPR. At AR = 1, $\lambda_{max} = 509$ nm, whilst at AR = 6, $\lambda_{max} = 821$ nm.

Kooij et al. [13] proposed a linear relationship between AR and λ_{max} where:

$$\lambda_{max}/nm = 415(\pm 6) + \eta 85.8(\pm 1.1) \tag{5.28}$$

and is plotted in Fig. 5.29, along with lower and upper error bounds. Alongside this λ_{max} has been plotted as taken from our measurements. Best fit calculations through our results yield a smaller gradient than the linear fit offered by Kooij, however the

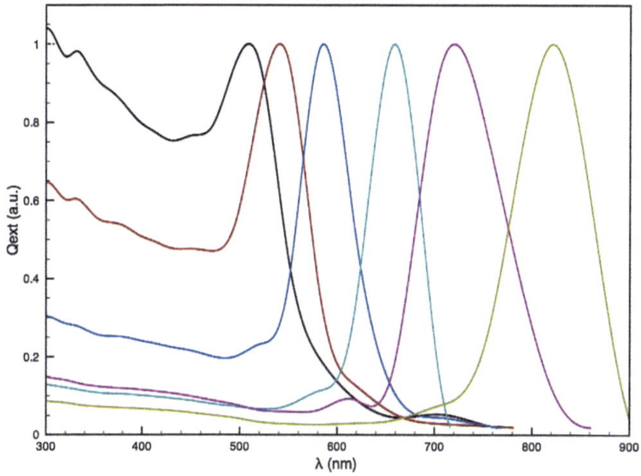

Fig. 5.28 Normalised extinction spectra for Au ellipsoids, with aspect ratios of 1 (*black*), 2 (*red*), 3 (*blue*), 4 (*green*), 5 (*purple*) and 6 (*olive*) plotted. Results were calculated using the original optical constants of Christy and Johnson [47], and then interpolated afterwards to give λ_{max} values. A *grey dotted line* is given to illustrate the maxima of λ_{max}

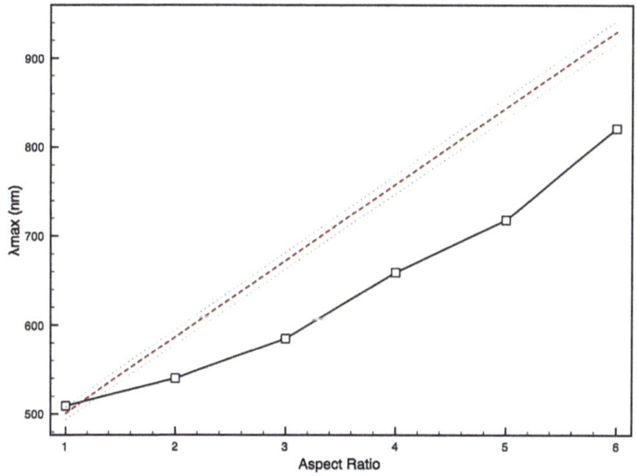

Fig. 5.29 Graph showing the relationship between AR and λ_{max} from our results (*black, solid line*), compared to a linear relationship (*red, dashed line*) given by Kooij et al. [13]. *Upper* and *lower* error bounds on this linear relationship are also given (*black, dotted lines*)

error bars of our λ_{max} values are large as they have been taken using post-calculation interpolation, and so this shows the requirement of using interpolated refractive index values at low energy (high λ) for calculations.

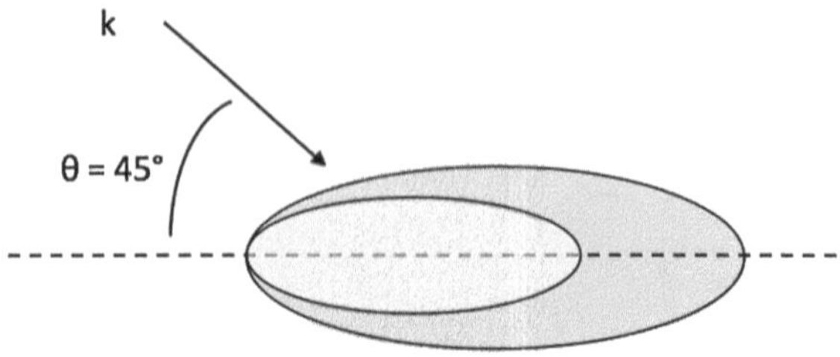

Fig. 5.30 A core-displaced ellipsoid model for a bimetallic core-shell particle. The core has been displaced along the (major) x-axis of the particle

5.4.2 Bimetallic Core-Shell Ellipsoids

5.4.2.1 Au-Pd

The addition of a Pd shell to an Au particle is known to have a strong effect on the damping of plasmon oscillations, as shown in Sect. 5.3.2 for core-shell spheres. However, in prolate ellipsoids the positioning of the core relative to the shell has not been previously investigated. A core-shell particle was designed with an ellipsoid core of AR = 2, with short axes of length $b_{core} = c_{core} = 2.0$ nm and long axes $a_{core} = \eta b_{core} = 4$ nm. This was encapsulated in an ellipsoid shell, illustrated in Fig. 5.6, with short axes $b_{shell} = c_{shell} = 1$ nm and long axis $a_{shell} = 2$ nm, giving overall parameters of $a = 6$ and $b = c = 3$ nm, thus maintaining AR = 2, and giving a core:shell volume stoichiometry of $\sim$1:3.5. Within the particle, the core was then displaced up to 1 nm along the x-axis, as illustrated in Fig. 5.30, and 0.5 nm along the y-axis (i.e. up to the particle edge in either direction).

Results are plotted in Fig. 5.31a for the extinction spectra of $Au_{core}Pd_{shell}$ and $Pd_{core}Au_{shell}$ prolate ellipsoids, with core-displacement along the x- and y-axis compared to a core-centred particle. For an Au ellipsoid of identical full size ($b = c = 3$ nm, $a = \eta b$), λ_{max} was calculated to be at 531 nm; this is a smaller size-induced blue-shift than for homogeneous Au particles studied in Sect. 5.4.1 where $\lambda_{max} = 540$ nm.

For $Au_{core}Pd_{shell}$, strong damping of all Au features is achieved by the Pd shell over the Au core ellipsoid, resulting in featureless spectra. However, for $Pd_{core}Au_{shell}$ a feature is identifiable at $\approx$ 520 nm. For the core-centred ellipsoid, $\lambda_{max} = 516$ nm; displacement of the Pd core along the y-axis slightly increases the intensity at λ_{max}, and slightly red-shifts the peak, whilst maximum core-displacement along the x-axis (2 nm) results in a more significant red-shift of λ_{max} to 524 nm, along with a further increase in intensity. As seen for spherical particles in Sect. 5.3.3.1, this can be attributed to the stronger Au LSPR resonance circumnavigating the Pd with greater ease when one end of the Pd elongated axis is displaced to the surface of the ellipsoid: this effect is amplified by the differing path lengths of dielectric material along the

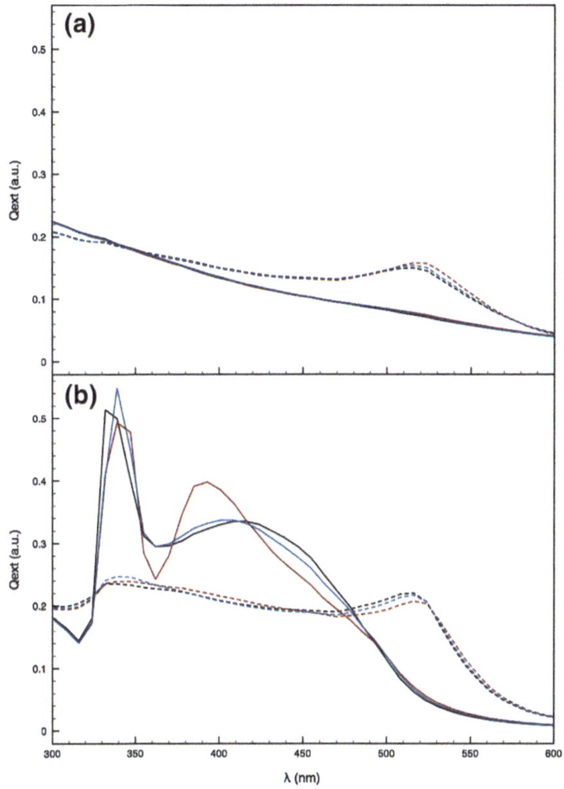

Fig. 5.31 Extinction spectra for core-shell elliptical particles with AR = 2: **a** $Au_{core}Pd_{shell}$ (*solid lines*) and the inverse $Pd_{core}Au_{shell}$ (*dashed lines*); **b** $Au_{core}Ag_{shell}$ (*solid lines*) and the inverse $Ag_{core}Au_{shell}$ (*dashed lines*). Core-centred geometries are given in *black*, whilst maximum core-displacement along the x-axis (2 nm) and y-axis (1 nm) are given in *red* and *blue*, respectively. Refractive indices have been interpolated from the literature. Reprinted with permission from Ref. [42]. Copyright 2012 American Chemical Society

major and minor axes, hence more significant spectral changes when the Pd core is displaced along the longer x-axis than the shorter y-axis.

5.4.2.2 Au-Ag

Results are plotted in Fig. 5.31b for the extinction spectra of $Au_{core}Ag_{shell}$ and $Ag_{core}Au_{shell}$ prolate ellipsoids, with core-displacement along the x- and y-axis compared to a core-centred particle. An Ag ellipsoids of dimensions similar to the core-shell particle ($b = c = 3$ nm, $a = \eta b$) gives resonances at 347 nm (TSPR) and $\lambda_{max} = 409$ nm (LSPR), consistent with previous experiments and calculations [28, 38].

For $Au_{core}Ag_{shell}$, the core-centred particle offers a strong TSPR resonance at 336 nm and a broad LSPR resonance at 416 nm, however no Au features are distinguishable. Whilst the red-shift of the LSPR relative to the Ag ellipsoid can be attributed to the influence of the Au core, the blue-shift of the TSPR relative to an Ag ellipsoid is less expected, though perhaps attributable to higher order resonances in the Au: quadrupolar resonances in Au exist at $\approx$ 330 nm, and at the low values of λ the dipolar approximation is perhaps at its limits. Maximum displacement of the Au core along the y-axis results in a red-shift of the TSPR to 339 nm, whilst the broad LSPR decreases in linewidth and blue-shifts to 409 nm. Displacing the Au core along the x-axis however results in a strong evolution of the Q_{ext} spectra, with the TSPR red-shifting further to ~343 nm, and the Ag LSPR feature peaking at 393 nm, with decreased FWHM compared to the pure Ag particle. The alignment of the TSPR with that of the pure Ag ellipsoid is expected here, as a greater bulk of Ag is now located together, however the LSPR is further blue-shifted than the pure Ag particle, implying that perhaps by having the Au dielectric concentrated on one side of the particle the AR is effectively reduced by the differing dielectrics.

Conversely, extinction spectra for $Ag_{core}Au_{shell}$ show the Au TSPR feature more prominently at ~510 nm than the Ag equivalents, though an Ag dipolar/Au quadrupolar shoulder is present at ~340 nm. In this arrangement, the Au feature at 516 nm (λ_{max}) is greatest when the core is centred; displacement along the x-axis reduces the intensity slightly, and red shifts λ_{max} to ~520 nm, whilst core-displacement along the y-axis shows little effect to λ_{max} compared to the core-centred ellipsoid, however it red shifts the shoulder at ~340 nm up to 350 nm. This implies that coupling of the Ag dipolar resonance with the Au quadrupolar has been disrupted when the long-side of the Ag core ellipsoid is moved towards the particle edge.

For both the $Au_{core}Ag_{shell}$ and $Ag_{core}Au_{shell}$ systems, no Au LSPR features appear in the extinction spectra, illustrating the dominant effect of the Ag in these systems even when it is not the greater quantity metal in the elliptical particle.

5.5 Cuboids

Experimental work synthesising $Au_{core}Pd_{shell}$ nanorods by Muche [60] found that whilst Au rods with controlled AR could be prepared (Fig. 5.32a), the addition of a thin Pd layer quenched all Au features in the UV-Visible absorption spectra. Imaging using STEM showed that the Au cores has been coated by the Pd shell forming cubic structures, with varying AR (Fig. 5.32b). The Au core in the cuboids was found to be between 48 and 78 nm in length, with a width of 18 nm; the Pd shell had a uniform thickness of 6 nm on all sides, forming an elongated cuboid with $2 \leq AR \leq 3$. Whilst considerably larger than the particles for which we have performed calculations, these dimensions are within the DDA criteria allowing computational theory to be applied.

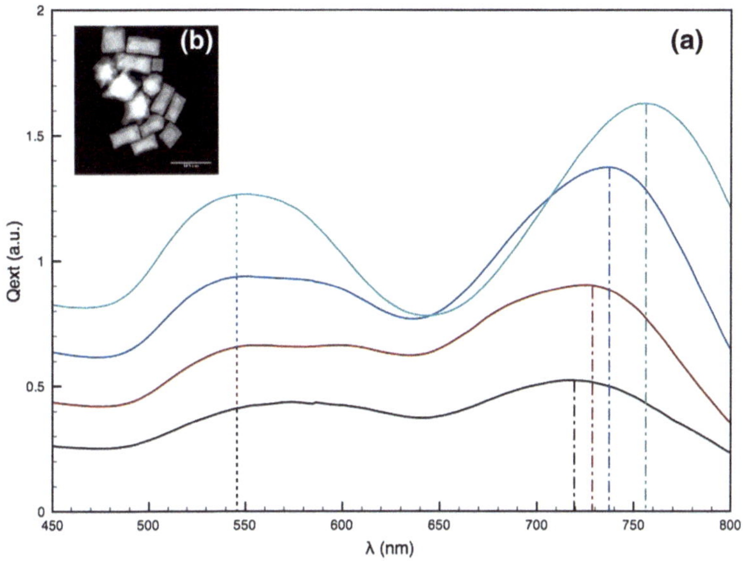

Fig. 5.32 Images reproduced from the experimental work by Muche [60]: **a** Extinction spectra for Au particles with increasing heating time, displayed in *black*, *red*, *blue* and *green*, respectively, resulting in an increase in intensity and red-shift of the LSPR (*dot-dashed line*). The TSPR remains relatively constant (*dotted line*). **b** STEM Image of particles with Pd coating; addition of a shell rendered the UV-Vis spectra featureless

5.5.1 Monometallic Au Cuboids

A series of Au cuboids were configured to replicate the experimental observations above, with a long-axis length of 48–78 nm, and a width of 18 nm. DDA calculations were performed on these geometries, with $\theta = 45°$ to ensure excitation of both the TSPR and LSPR, and the results are plotted in Fig. 5.33. Whilst for Au cubes a spectral feature is expected at 600 nm [26, 61], we find that extension of one length parameter results in decoupling of the TSPR and LSPR features, as seen for the transition between spheroids and ellipsoids. For increasing cuboid length, and thus increasing AR, the LSPR is found to red-shift linearly and increase in intensity, akin to the experimental spectra (Fig. 5.32); a small residual TSPR resonance is present at ~520 nm throughout.

In Fig. 5.33 an inset plots AR against λ_{max}, which we have compared to Kooij's relationship for an Au ellipsoid with varying AR (Eq. 5.28); we found that by altering the gradient to this relationship to:

$$\lambda_{max}/nm = 415(\pm 6) + \eta 69.8(\pm 1.1) \tag{5.29}$$

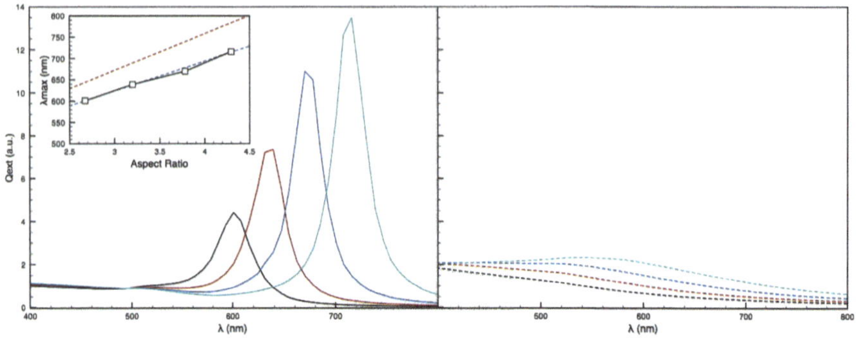

Fig. 5.33 Extinction spectra for cuboid particles. *Left* Au cuboid, with AR of 2.67 (*black*), 3.2 (*red*), 3.78 (*blue*) and 4.3 (*green*) plotted. *Inset* is a plot of AR against λ_{max} to illustrate the shift with increasing AR (*black*, *solid line*). Linear fits are also given as suggested by Kooij et al. for ellipsoids (Eq. 5.28; *red*, *dotted line*), and a corrected function with a gentler gradient (Eq. 5.29; *blue*, *dotted line*). *Right* Au cuboid with Pd shell of thickness 6 nm, corresponding to $2 \leq \text{AR} \leq 3$. ARs of 2 (*black*), 2.33 (*red*), 2.66 (*blue*) and 3 (*green*) are given. Refractive indices have been interpolated from experimental values

an excellent agreement with the calculated values of λ_{max} was achieved. λ_{max} for cuboids is at a longer wavelength than seen for ellipsoids, due to the larger volume of Au dielectric for the LSPR to oscillate across.

5.5.2 *Bimetallic Core-Shell Cuboids*

5.5.2.1 Au-Pd

A Pd coating, of 6 nm thickness, was added to the Au cores from Sect. 5.5.1, to form $Au_{core}Pd_{shell}$ cuboids with $2 \leq AR \leq 3$, similar to those seen experimentally (Fig. 5.32). The results of DDA calculations for these structures are plotted in Fig. 5.33: at all AR the prominent Au LSPR features have been absorbed, being replaced by very gentle spurious Pd shoulders. The absorption of the Au features at λ_{max} is not surprising, given that thinner shells of Pd have quenched Au features previously in our calculations (Sect. 5.3.2.1), but corroboration with experiment is achieved.

5.6 Cylinders

Nanorods are of continued interest to the nano-optic community, and the synthesis of bimetallic particles is challenging and complex as is demonstrated by the work of Muche [60]. Likewise, to model nanorods many different geometries can be used

[18, 26]: ellipsoids and cuboids we have already seen in this chapter. We complete this work by performing DDA calculations on Au cylinders, and then creatively adapting these with a secondary group X metal to show the hypothetical tuning of the Au spectral features within the range $650 \leq \lambda \leq 850$ nm.

5.6.1 Monometallic Au Cylinders

Flat-ended Au cylinders, with lengths of 40 and 50 nm and $r = 5$ nm (i.e. AR = 4 and 5, respectively), were prepared and subjected to DDA calculations, with an incident angle of 45°. Another hemispherically capped cylinder with length = 50 nm and AR = 5 was also used to verify the importance of rod endings to the position of λ_{max}; schematic structures are shown in Fig. 5.34, along with schematics for proposed alterations. The results are displayed in Fig. 5.35.

For the Au flat-ended cylinders we find increases in AR red-shift λ_{max}, as seen for other elongated structures. The ratio of change in λ_{max} is very similar to that seen for the cuboid structures in Sect. 5.5.1, with the adapted linear relationship in Eq. 5.29 correct to ±10 nm. Adding hemispherical endings to the AR = 4 cylinder, thus increasing the AR to 5, gives a λ_{max} value blue-shifted from the flat-ended cylindrical equivalent; this can be likened to the blue-shift of λ_{max} for ellipsoidal

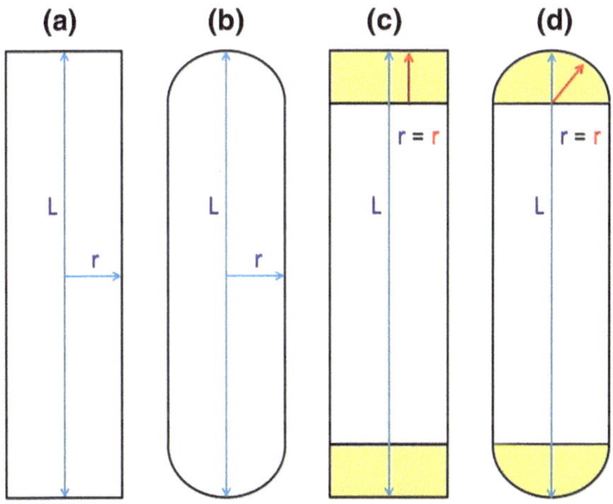

Fig. 5.34 2D schematics of cylindrical structures: **a** Homogeneous flat-ended cylinder, with length L and radius r, as marked by the *blue arrows*. **b** Homogeneous hemispherical capped cylinder, with identical AR to (**a**). **c** Heterogeneous flat-ended cylinder, equivalent to (**a**) but flat-ended cylindrical capping of depth r on each end by a second material (*yellow*). **d** Heterogeneous hemispherically capped cylinder, equivalent to (**b**) but with endings replaced with the dielectric of a second material (*yellow*) to depth r. In both (**c**) and (**d**) the core particle has the same length, equal to $L - 2r$

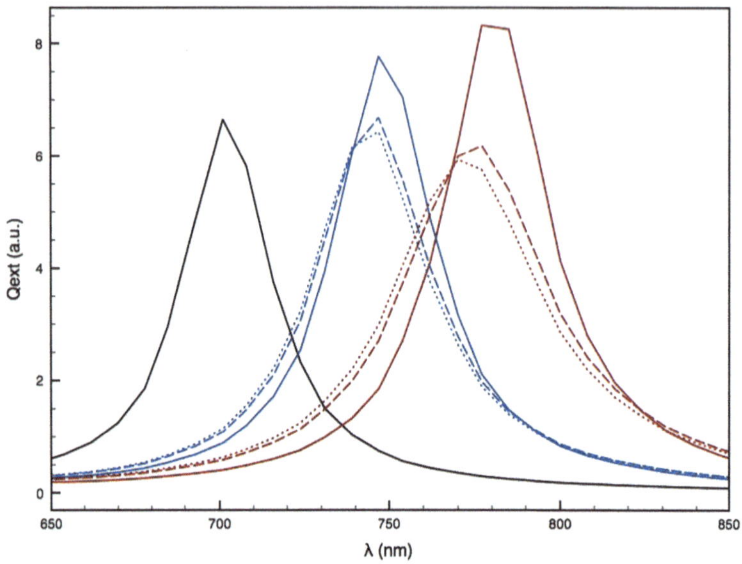

Fig. 5.35 Extinction spectra for: (i) Au cylinders with flat-ends with AR of 4 (*black*) and 5 (*red*), and hemi-spherically capped cylinder with AR of 5 (*blue*), given in *solid lines*. (ii) Au cylinders with Pd endings of AR = 5, plotted with *dashed lines*, showing Q_{ext} for flat-ended (*red*) and hemi-spherically (*blue*) capped cylinders. (iii) As with (ii), except dielectric material at the ends has been changed to that of Pt; plotted with *dotted lines*

particles compared to cuboids, with reduced full length dielectric volume available to propagate through.

5.6.2 Bimetallic Core-Shell Cuboids

5.6.2.1 Au-Pd

Addition of a secondary shell metal to an Au core has shown to change the extinction spectra thus far in our work. Taking a flat-ended AR = 4 cylinder, Pd hemispherical and flat-ended additions were added on to the ends of the cylinders, taking the AR up to 5 (Fig. 5.34). The results are plotted in Fig. 5.35.

Addition of the Pd endings is shown to dampen the LSPR response of the flat-ended and hemispherical ended cylinders, compared to the pure Au equivalent, and is coupled with an increase in the FWHM. Whilst a shift in λ_{max} cannot be identified without more detailed calculations, a slight blue-shift of plasmon attenuation is noted for both types of "cappings".

5.6.2.2 Au-Pt

As a comparison we replaced the Pd nanorod endings with the dielectric of Pt, another group X element. The results are plotted in Fig. 5.35, and whilst the same observations as for Pd-capping holds true (damping, increased FWHM) a perceptible blue-shift in λ_{max} is found for the flat-ended capped-cylinder, relative to the Au equivalent with AR = 5. Pt capping growth on the end of Au rods has been seen experimentally and tested computationally [62], and this type of bimetallic synthesis offers potential to further tailor the extinction spectra of nanoparticles.

5.7 Conclusions

We have investigated the optical extinction properties of Au nanoparticles using the DDA, showing that geometry heavily influences the position of λ_{max}. For spheres, $\lambda_{max} = 508$ nm, whilst for ellipsoids, cuboids and cylinders the TSPR remains at the same value of λ, however λ_{max} is linked with the LSPR, which red-shifts with a dependency on the aspect ratio of the nanorod for the size of the red-shift. Increasing size results in a greater intensity of Q_{ext}. We have also shown that coupling of Au nanospheres results in a red-shift in λ_{max}, akin to the response of a nanorod with increasing AR.

Core-shell Au-M nanoparticles (M = Pd, Pt, Ag) have also been investigated, with attention paid to the effect of varying the position of the core within the particle. For core-centred particles, Pd and Ag shells are found to damp Au features rapidly. Au shells on Pd and Ag cores require thick layers to dampen core features; for Ag this is still difficult. Displacement of the core within the Au-M structure results in an increase in intensity of Q_{ext} at λ_{max}, with the greatest effect seen when the core is touching the particle surface. This is due to the increased concentration of dielectric volume in close proximity, exemplified by the response of Janus particles. The intensity of response of $Au_{core}Pd_{shell}$ particles is found to be orientation dependent also.

For conjoined $Au_{core}Pd_{shell}$ nanospheres we have identified previously unreported optical spectra phenomena. Furthermore, we have shown that different extents of overlap between $Au_{core}Pd_{shell}$ nanospheres results in the evolution of the extinction spectral features from those characteristic of a sphere towards those of rod shaped geometries. Increasing the refractive index of the surrounding medium result in strong red shifts and broadening of prominent peaks, with the appearance of a residual TSPR feature when Au cores are not in contact. Comparison of our results with previous work on Au nanorods illustrates the progression from $Au_{core}Pd_{shell}$ spheres towards more rod like attributes in the extinction spectrum.

Core positioning also plays a key role in the position of λ_{max} in core-shell Au-M ellipsoids, with AR of 2 and higher. Whilst changes to the Au-Pd spectra are slight, the movement of the elliptical core in the $Au_{core}Ag_{shell}$ results in red-shifting of λ_{max} and blue-shifting of the LSPR feature, with reduced FWHM in the latter. For

$Au_{core}Pd_{shell}$ cuboids we find that Au spectral features are damped by the Pd shell, as found in experiments.

Finally, we investigated the effect of adding different metals to the end of Au cylinders. Relative to homogeneous Au cylinders, we find that adding Pt and Pd to the end of the cylinders, whilst maintaining a constant AR, results in damping of λ_{max} at the LSPR. In the case of Pt we also see slight blue-shift of λ_{max}.

Overall, we have shown that the optical response of nanoparticles is finely tunable with different geometries and constituent metals. Whilst the accuracy at which experiment currently operates may not be suitable for identifying some of the subtle optical changes we have identified as possible, there is plenty of scope for fine-tuning in further work. The position of λ_{max} is particular important, and further investigations could be usefully performed on conjoined core-shell particles with off-centred cores. It would also be of interest to expand our investigations to include copper, which is more suitable industrially for use in nano-optics due to its lower market cost.

References

1. R.H. Ritchie, Plasma losses by fast electrons in thin films. Phys. Rev. **106**(5), 874–881 (1957)
2. E.A. Stern, R.A. Ferrell, Surface plasma oscillations of a degenerate electron gas. Phys. Rev. **120**(1), 130–136 (1960)
3. Z.Y. Li, N.P. Young, M. Di Vece, S. Palomba, R.E. Palmer, A.L. Bleloch, B.C. Curley, R.L. Johnston, J. Jiang, J. Yuan, Three-dimensional atomic-scale structure of size-selected gold nanoclusters. Nature **451**(7174), 46–48 (2008)
4. H.A. Atwater, A. Polman, Plasmonics for improved photovoltaic devices. Nat. Mat. **9**, 205–213 (2010)
5. F. Baletto, R. Ferrando, Structural properties of nanoclusters: energetic, thermodynamic, and kinetic effects. Rev. Mod. Phys. **77**, 371–423 (2005)
6. U. Kreibig, M. Vollmer, *Optical Properties of Metal Clusters* (Springer, Berlin, 1995)
7. G. Mie, Beiträge zur Optik trüber Medien, speziell kolloidaler Metallösungen. Annal. Phys. **25**, 377 (1908)
8. J.R. Wait, Electromagentic scattering from a radially inhomogeneous sphere. J. Appl. Sci. Res. B **10**(5), 441–450 (1963)
9. R. Gans, Über die Form ultramikroskopischer Goldteilchen. Annal. Phys. **342**, 881 (1912)
10. J.R. Wait, Scattering of a plane wave from a circular dielectric cylinder at oblique incidence. Can. J. Phys. **33**(5), 189–195 (1955)
11. B.T. Draine, P.J. Flatau, Discrete-dipole approximation for scattering calculations. J. Opt. Soc. Am. A **11**(4), 1491–1499 (1994)
12. L.B. Scaffardi, N. Pellegri, O. de Sanctis, J.O. Tocho, Sizing gold nanoparticles by optical extinction spectroscopy. Nanotechnology **16**(1), 158–163 (2005)
13. E.S. Kooij, B. Poelsema, Shape and size effects in the optical properties of metallic nanorods. Phys. Chem. Chem. Phys. **8**, 3349–3357 (2006)
14. S.M. Morton, D.W. Silverstein, L. Jensen, Theoretical studies of plasmonics using electronic structure methods. Chem. Rev. **111**, 3962–3994 (2011)
15. M.M. Alvarez, J.T. Khoury, T.G. Schaaff, M.N. Shafigullin, I. Vezmar, R.L. Whetten, Optical absorption spectra of nanocrystal gold molecules. J. Phys. Chem. B **101**(19), 3706–3712 (1997)
16. M.L. Sandrock, C.A. Foss, Synthesis and linear optical properties of nanoscopic gold particle pair structures. J. Phys. Chem. B **103**, 11398–11406 (1999)

17. A. Brioude, X.C. Jiang, M.P. Pileni, Optical properties of gold nanorods: DDA simulations supported by experiments. J. Phys. Chem. B **109**(27), 13138–13142 (2005)
18. S.W. Prescott, P. Mulvaney, Gold nanorod extinction spectra. J. Appl. Phys. **99**(12), 123504 (2006)
19. K.-S. Lee, M.A. El-Sayed, Dependence of the enhanced optical scattering efficiency relative to that of absorption for gold metal nanorods on aspect ratio, size, end-cap shape, and medium refractive index. J. Phys. Chem. B **109**(43), 20331–20338 (2005)
20. E.M. Larsson, J. Alegret, M. Käll, D.S. Sutherland, Sensing characteristics of nir localized surface plasmon resonances in gold nanorings for application as ultrasensitive biosensors. Nano Lett. **7**(5), 1256–1263 (2007)
21. J. Chen, B. Wiley, Z.-Y. Li, D. Campbell, F. Saeki, H. Cang, L. Au, J. Lee, X. Li, Y. Xia, Gold nanocages: Engineering their structure for biomedical applications. Adv. Mater. **17**(18), 2255–2261 (2005)
22. P.K. Jain, S. Eustis, M.A. El-Sayed, Plasmon coupling in nanorod assemblies: optical absorption, discrete dipole approximation simulation, and exciton-coupling model. J. Phys. Chem. B **110**, 18243–18253 (2006)
23. S.K. Ghosh, T. Pal, Interparticle coupling effect on the surface plasmon resonance of gold nanoparticles: from theory to applications. Chem. Rev. **107**, 4797–4862 (2007)
24. C. Noguez, Surface plasmons on metal nanoparticles: the influence of shape and physical environment. J. Phys. Chem. C **111**(10), 3806–3819 (2007)
25. I.O. Sosa, C. Noguez, R. Barrera, Optical properties of metal nanoparticles with arbitrary shapes. J. Phys. Chem. B **107**(3), 6269–6275 (2003)
26. A.L. Gonzalez, J.A. Reyes-Esqueda, C. Noguez, Optical properties of elongated noble metal nanoparticles. J. Phys. Chem. C **112**(19), 7356–7362 (2008)
27. R. Ferrando, J. Jellinek, R.L. Johnston, Nanoalloys: From theory to applications of alloy clusters and nanoparticles. Chem. Rev. **108**(3), 845–910 (2008)
28. M.B. Cortie, A.M. McDonagh, Synthesis and optical properties of hybrid and alloy plasmonic nanoparticles. Chem. Rev. **111**(6), 3713–3735 (2011)
29. F. Liu, D. Wechsler, P. Zhang, Alloy-structure-dependent electronic behavior and surface properties of Au Pd nanoparticles. Chem. Phys. Lett. **461**, 254–259 (2008)
30. H. Zhang, T. Watanabe, M. Okumura, M. Haruta, N. Toshima, Catalytically highly active top gold atom on palladium nanocluster. Nat. Mater. **11**, 49–52 (2011)
31. C. Kan, W. Cai, C. Li, L. Zhang, H. Hofmeister, Ultrasonic synthesis and optical properties of Au/Pd bimetallic nanoparticles in ethylene glycol. J. Phys. D: Appl. Phys. **36**(13), 1609–1614 (2003)
32. N. Toshima, Core/shell-structured bimetallic nanocluster catalysts for visible-light-induced electron transfer. Pure Appl. Chem. **72**, 317–325 (2000)
33. Y. Xiang, X. Wu, D. Liu, X. Jiang, W. Chu, Z. Li, Y. Ma, W. Zhou, S. Xie, Formation of rectangularly shaped Pd/Au bimetallic nanorods: evidence for competing growth of the Pd shell between the 110 and 100 side facets of Au nanorods. Nano Lett. **6**(10), 2290–2294 (2006)
34. J.H. Song, F. Kim, D. Kim, P. Yang, Crystal overgrowth on gold nanorods: tuning the Shape, Facet, Aspect Ratio, and composition of the nanorods. Chemistry **11**(3), 910–916 (2005)
35. L.M. Liz-Marzán, Tailoring surface plasmons through the morphology and assembly of metal nanoparticles. Langmuir **22**(1), 32–41 (2006)
36. Z.Y. Li, J.P. Wilcoxon, F. Yin, Y. Chen, R.E. Palmer, R.L. Johnston, Structures and optical properties of 4–5 nm bimetallic AgAu nanoparticles. Faraday Discuss. **138**, 363 (2008)
37. R. Taylor, *The Optical Properties of Au Nanoparticles*, Master's thesis, School of Physics and Astronomy (University of Birmingham, Birmingham, 2009)
38. K.L. Kelly, E. Coronado, L.L. Zhao, G.C. Schatz, The optical properties of metal nanoparticles: The influence of size, shape, and dielectric environment. J. Phys. Chem. B **107**(3), 668–677 (2003)
39. V.M. Rotello, *Nanoparticles: Building Blocks for Nanotechnology* (Springer Science+Business Media, Inc., USA, 2004)

40. G.V. Hartland, Optical studies of dynamics in noble metal nanostructures. Chem. Rev. **111**, 3858–3887 (2011)
41. J. Pérez-Juste, I. Pastoriza-Santos, L.M. Liz-Marzán, P. Mulvaney, Gold nanorods: synthesis, characterization and applications. Coordin. Chem. Rev. **249**(17–18), 1870–1901 (2005)
42. A.J. Logsdail, R.L. Johnston, Predicting the optical properties of core-shell and janus segregated au-m nanoparticles (m = ag, pd). J. Phys. Chem. C **116**(44), 23616–23628 (2012)
43. W.L. Barnes, A. Dereux, T.W. Ebbesen, Surface plasmon subwavelength optics. Nature **424**, 824–830 (2003)
44. E.M. Purcell, C.R. Pennypacker, Astrophys. J. **186**, 705–714 (1973)
45. M.A. Yurkin, A.G. Hoekstra, The discrete dipole approximation: an overview and recent developments. J. Quant. Spectrosc. Radia. Transf. **106**, 558–589 (2007)
46. B.T. Draine, J.J. Goodman, Beyond Clausius-Mossotti: Wave propagation on a polarizable point lattice and the discrete dipole approximation. Astrophys. J. **405**, 685 (1993)
47. P.B. Johnson, R.W. Christy, Optical constants of the noble metals. Phys. Rev. B **6**(12), 4370–4379 (1972)
48. P.B. Johnson, R.W. Christy, Optical constants of transition metals: Ti, V, Cr, Mn, Fe, Co., Ni, and Pd. Phys. Rev. B **9**(12), 5056–5070 (1974)
49. E.D. Palik, *Handbook of Optical Constants of Solids* (Academic Press, New York, 1985)
50. http://www.bear.bham.ac.uk/bluebear/
51. J.C. Slater, Atomic radii in crystals. J. Chem. Phys. **41**(10), 3199 (1964)
52. A. Walther, A.H.E. Muller, Janus particles. Soft Matter **4**, 663–668 (2008)
53. J. Phillips, *Simulation of the Optical Properties of Ag-Au Nanoparticles*, Master's thesis, School of Physics and Astronomy (University of Birmingham, Birmingham, 2008)
54. N.J. Cookson, *Preparation and Characterisation of Bimetallic Core-Shell Nanoparticles*, Master's thesis, School of Chemistry (University of Birmingham, Birmingham, 2009)
55. A.J. Logsdail, N.J. Cookson, S.L. Horswell Z.W., Wang, Z.Y. Li, R.L. Johnston, Theoretical and experimental studies of the optical properties of conjoined gold palladium nanospheres. J. Phys. Chem. C **114**(49), 21247–21251 (2010)
56. I. Romero, J. Aizpurua, G.W. Bryant, F.J. García De Abajo, Plasmons in nearly touching metallic nanoparticles: singular response in the limit of touching dimers. Opt. Express **14**(21), 9988–9999 (2006)
57. L.S. Slaughter, W.-S. Chang, P. Swanglap, A. Tcherniak, B.P. Khanal, E.R. Zubarev, S. Link, Single-particle spectroscopy of gold nanorods beyond the quasi-static limit: Varying the width at constant aspect ratio. J. Phys. Chem. C **114**(11), 4934–4938 (2010)
58. B. Yan, Y. Yang, Y. Wang, Comment on simulation of the optical absorption spectra of gold nanorods as a function of their aspect ratio and the effect of the medium dielectric constant. J. Phys. Chem. B **107**, 9159 (2003)
59. Z. Zhong, S. Patskovskyy, P. Bouvrette, J.H.T. Luong, A. Gedanken, The surface chemistry of Au Colloids and their interactions with functional amino acids. Phys. Rev. B **108**, 4046–4052 (2004)
60. T.A. Muche, *Synthesis, Characterisation and Applications of Bimetallic Nanoparticles*, Master's thesis, School of Chemistry (University of Birmingham, Birmingham, 2010)
61. W. Kohn, L.J. Sham, Self-consistent equations including exchange and correlation effects. Phys. Rev. **140**, A1133–A1138 (1965)
62. A.J. Cohen, P. Mori-Sánchez, W. Yang, Challenges for density functional theory. Chem, Rev. **112**(1), 289–320 (2012)

Chapter 6
Concluding Remarks

6.1 Conclusions

In this research we have investigated the structures, characterisation and optical responses of monometallic and bimetallic nanoclusters, focusing on gold particles, and gold-containing nanoalloys, due to their optical and catalytic properties in the nano-regime. We have coupled our work, where possible, with the experimental observations of others to validate our results.

We began in Chap. 2 by looking at the structural preference of high-symmetry Au_N, Pd_N and $(AuPd)_N$ nanoclusters, up to $N = 6{,}000$. For Pd_N clusters we found preference to the high-coordination icosahedral structure (Ih) for $N < 1,000$ when compared to cuboctahedra (CO) and Ino-decahedra (I-Dh), two other high-symmetry 12-vertex structures. However, truncated-octahedra (TO) with reduced (100) faces and Marks-decahedra (M-Dh) of similar size prove to have lower excess energies. Au_N clusters favour FCC structures at lower values of N than for Pd_N: the parameters of Cleri and Rosato [1] give a transition at $N \approx 650$, whilst the parameters of Baletto et al. [2] give a lower value. As for Pd_N, TO and M-Dh structures with reduced (100) exposure prove structurally favourable for Au_N.

Bimetallic $(AuPd)_N$ clusters show thermodynamic preference for a $(Pd_{core}Au_{shell})_N$ arrangement. Monolayers of Au prove most energetically favourable, compared to the pure monometallic clusters. Ih structural motifs are more favourable for $(Pd_{core}Au_{shell})_N$ clusters at low nuclearities ($N \lesssim 400$), though this is dependent on the size of the Pd core. $(Au_{core}Pd_{shell})_N$ clusters are not energetically favourable with thin coatings of Pd, however as the shell coating thickens so the stability improves: the parameters of Cleri and Rosato give prominence to this. Overall, a particularly strong tendency towards core/shell segregation is observed for the parameters of Cleri and Rosato [1], and not so much for the parameters of Baletto et al. [2], in agreement with previous work on smaller $(AuPd)_N$ clusters [3, 4].

In Chap. 3 we discussed, developed and implemented a HAADF-STEM model, which was coupled with a genetic algorithm (GA), using statistical methods to compare simulated and experimental HAADF-STEM images. Of the comparative

A. J. Logsdail, *Computational Characterisation of Gold Nanocluster Structures*, Springer Theses, DOI: 10.1007/978-3-319-01493-7_6,

methods implemented, least squares fitting (LSF) proves to be more susceptible to the configuration of an experimental image: data needed to be centralised, with intensities normalised. Covariance has also been implemented, and a residual ($\sigma_{\theta,\phi,\psi}$) has been used to gauge the relative suitability of a theoretical structure's orientation, dependent on the Tait-Bryan angles θ, ϕ and ψ, to alternative orientations, when tested against a predefined solution.

Initially, linear searches were conducted of the function landscape when comparing monolayer-protected Au_{38} clusters with an experimental image. From this we were able to identify methodological improvements, such as using minimisation to improve our search efficiency. Local and global minimisation were implemented, tested and parameterised. We show that increasing frequency of the genetic operators improves search efficiency: mutation of population parents proves to be the most effective at improving convergence rates, though it does increase function evaluations (FEs) as well. Whilst Darwinian genetic algorithms (D-GAs) improve search results compared to a random, non-minimising search, implementation of a computationally non-intensive local minimisation process, such as Powell's method, to form a Lamarckian genetic algorithm (L-GA) greatly improves search efficiency. For D-GAs the quantity of calls to the search history (historical evaluations, HEs) does not significantly increase with genetic frequency, but the same is not true for L-GAs.

However, random locally-minimising searches prove more efficient than L-GAs by an order of magnitude. The most efficient local minimisation method proves to be Powell's method, where an additional search vector improves the analytical minimisation process; in some circumstances this may be because the search vector allows "tunnelling" through an otherwise impassable function barrier.

We have also tested our STEM-GA for its resilience to background noise, for which a threshold was found at 4convergence to the correct orientation decreases. Finally, the use of parallel computing has been incorporated into our program to improve "real" search speeds, harnessing both shared memory (OMP) and non-shared memory (MPI) parallelisation techniques.

Chapter 4 details calculations of the adsorption of the T_d-symmetry Au_{16} cluster on to two layers of highly-ordered pyrolytic graphite (HOPG) using density functional theory (DFT). Whilst bonding to a pristine two-layer graphite surface (GR2) proves chemically weak, the introduction of surface defects increases the adsorption energy (E_a) at the cost of distortion of the Au_{16} cluster, due to the chemical interactions between the gold atoms and "dangling" carbon bonds around the vacant sites. It is noted that Au atoms prefer to adsorb in the carbon plane if possible, for which a vacancy of two adjacent C atoms is necessary. This proves detrimental to potential use in catalysis of CO oxidation, with distortion of the Au_{16} cluster leading to low levels of charge transfer to the cluster ($\Delta q \rightarrow 0$). ELF calculations show that the AuC bond is delocalised in nature, unlike covalent CC bonds.

Further testing with Van der Waals-corrected exchange-correlation functionals gives increased E_a and slightly shorter Au-X distances for deposition on GR2, but for a surface with two defects (GR2-2h) the effects are reversed due to the rearrangement of long-range "aurophilic" interactions within the Au_{16} cluster.

The potential catalytic uses of Au_{16} deposited on HOPG, e.g. CO oxidation, were investigation with the soft-landing of an O_2 molecule on the gold cluster. Whilst the initial charge on the Au_{16} cluster was promising, charge transfer on to the O_2 molecule to create a superoxo-state was difficult, with success only when the O_2 molecule interacts with more than one Au atom. This has since been investigated for Au_{16} deposited on graphane by Chen et al. [5].

Finally, Chap. 5 looked at the optical extinction properties of Au nanoparticles, and nanoalloy derivatives, using the discrete dipole approximation (DDA). For spheres, $\lambda_{max} = 508$ nm, whilst for ellipsoids, cuboids and cylinders the transverse surface plasmon resonance (TSPR) remains at the same value of λ, however λ_{max} is linked with the longitudinal surface plasmon resonance (LSPR), which displays a red-shifts depending on the aspect ratio of the nanorod. The LSPR is also shown to red-shift with coupled Au spheres, and increased volume gives increased intensity on the extinction spectra.

In core-shell Au-M nanoparticles (M = Pd, Pt, Ag) the shell thickness and material has the largest influence on the extinction spectra, with the location of the core (centred or off-centred) having a more subtle effect on intensity. Pd and Ag shells are found to damp Au extinction features rapidly, whilst Au shells on Pd and Ag cores require thick layers to dampen core features. For conjoined $Au_{core}Pd_{shell}$ particles we have investigated the effect of differing degrees of particle overlap, with the extinction spectra evolving from that of a sphere to that of a nanorod as overlap increases. The refractive index of the surrounding medium was investigated, giving red-shifts to extinction features and broadening of prominent peaks.

We looked at off-centred cores in core-shell Au-M ellipsoids with an aspect ratio (AR) of 2 or higher. For M = Pd the effects are slight, however they are much more pronounced for M = Ag: the movement of the elliptical core in the $Au_{core}Ag_{shell}$ results in red-shifting of λ_{max} and blue-shfting of the LSPR feature, with reduced FWHM in the latter. Lastly, we looked at the effect of selectively adding metal to the end of Au cylinders, which resulted in damping of extinction features relative to pure Au rods of the same aspect ratio.

6.2 Future Work

Given the breadth of work studied in this thesis, there are many avenues available for further investigation: both in increasing the diversity of elements and structures on which calculations are performed (Chaps. 2, 4, and 5) and also in the further development of models and search methods (Chap. 3).

Exploration of other high- and low-symmetry homotops for bimetallic $(AuPd)_N$ clusters is necessary, following the work in Chap. 2, as well as investigation of the stability of nanoclusters with off-centred cores. This work could be expanded to look at other coinage metals, i.e. Ag and Pt, so as to fit in with our optical calculations. Similarly, following on from the work in Chap. 5, it would be beneficial to look at the optical spectra for conjoined core-shell particles with off-centred cores, as well as expanding our calculations to include copper, an industrially suitable metal for applications in nano-optics. The ability to fine tune the position of λ_{max} and the intensity of the extinction response would be of great commercial use, e.g. biomedical applications [6].

The work in Chap. 3 offers the opportunity for model and program development, following on from the preliminary work which has been conducted. Whilst the model and search algorithm have been developed successfully, improvements could be made in our searches to take into consideration cluster symmetry in order to decrease search space. Also, our kinematic model could be further improved by taking into consideration the mobility of atoms, due to vibrations or structural rearrangement, as seen in clusters under the electron beam (particularly small nanoclusters [7]). Several methods have been proposed, of which the incorporation of a Debye-Waller factor into the model seems more appropriate. The probe could be improved with the use of an Airy function [8], whilst for larger nanoclusters with periodic packing it would be better to use Fourier transforms to ease the process of structure identification. Fast Fourier transforms (FFTs) are implemented in our code, but are yet to be fully tested.

Finally, the work in Chap. 4 has clear potential for future calculations into the effect of defect size, and of either surface doping or system intercalation. This would clarify whether Au_{16} particles offer any potential catalytic applications. Furthermore, calculations could be expanded to look at small clusters of other precious metals, such as Pd, and bimetallic derivatives (as investigated in Chap. 2). With regards current applications of the Au_{16} cluster, further study should look at O_2 dimers interacting with 2 (or more) Au atoms, and then CO oxidation can be investigated using methods which explore reaction pathways, such as the nudged elastic-band method [9].

References

1. F. Cleri, V. Rosato, Tight-binding potentials for transition metals and alloys. Phys. Rev. B **48**(1), 22–33 (1993)
2. F. Baletto, R. Ferrando, A. Fortunelli, F. Montalenti, C. Mottet, Crossover among structural motifs in transition and noble-metal clusters. J. Chem. Phys. **116**(9), 3856–3863 (2002)
3. F. Pittaway, L.O. Paz-Borbón, R.L. Johnston, H. Arslan, R. Ferrando, C. Mottet, G. Barcaro, A. Fortunelli, Theoretical studies of palladium-gold nanoclusters: Pd-Au clusters with up to 50 atoms. J. Phys. Chem. C **113**, 9141–9152 (2009)
4. R. Ismail, R.L. Johnston, Investigation of the structures and chemical ordering of small pd-au clusters as a function of composition and potential parameterisation. Phys. Chem. Chem. Phys. **12**(30), 8607–8619 (2010)
5. G. Chen, S.J. Li, Y. Su, V. Wang, H. Mizuseki, Y. Kawazoe, Improved stability and catalytic properties of au16 cluster supported on graphane. J. Phys. Chem. C **115**(41), 20168–20174 (2011)
6. J. Chen, B. Wiley, Z.-Y. Li, D. Campbell, F. Saeki, H. Cang, L. Au, J. Lee, X. Li, Y. Xia, Gold nanocages: Engineering their structure for biomedical applications. Adv. Mater. **17**(18), 2255–2261 (2005)
7. Z. Wang, O. Toikkanen, B.M. Quinn, R.E. Palmer, Real-space observation of prolate monolayer-protected Au_{38} clusters using aberration-corrected scanning transmission electron microscopy. Small **7**(11), 1542–1545 (2011)

8. S.J. Pennycook, P.D. Nellist, *Scanning Transmission Electron Microscopy: Imaging and Analysis* (Springer, New York, 2011)
9. D. Sheppard, R. Terrell, G. Henkelman, Optimization methods for finding minimum energy paths. J. Chem. Phys. **128**(13), 134106 (2008)

Appendix A
Plot of Gupta Potential and Cutoff

See Fig. A.1.

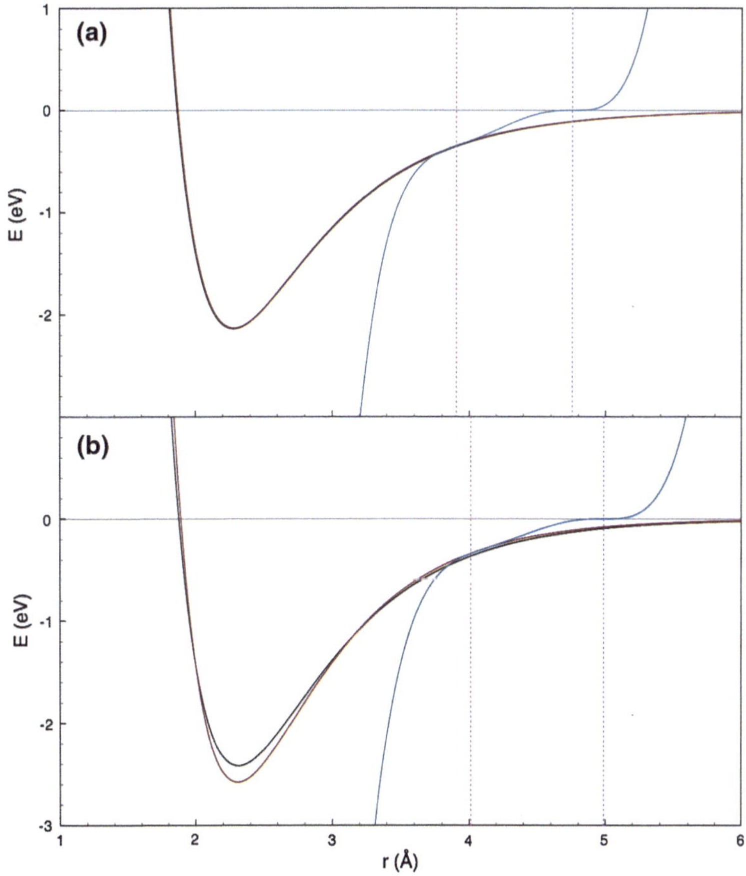

Fig. A.1 Plot of the Gupta Potential function, with different parameters and cutoff in each graph for **a** Pd and **b** Au. *Black lines* represent the parameters of Cleri and Rosato [1], *red lines* represent the parameters of Baletto et al. [2], and the *blue lines* show the polynomial cutoff, with C_s and C_e shown by *pink* and *purple dashed lines*, respectively

A. J. Logsdail, *Computational Characterisation of Gold Nanocluster Structures*, Springer Theses, DOI: 10.1007/978-3-319-01493-7,

References

1. F. Cleri, V. Rosato, Tight-binding potentials for transition metals and alloys. Phys. Rev. B **48**(1), 22–33 (1993)
2. F. Baletto, R. Ferrando, A. Fortunelli, F. Montalenti, C. Mottet, Crossover among structural motifs in transition and noble-metal clusters. J. Chem. Phys. **116**(9), 3856–3863 (2002)

Appendix B
Data from the Development of a Genetic Algorithm to Characterise Scanning Transmission Electron Microscope Images

B.1 Thiolate-Protected Au_{38} with Ligands Removed: $\alpha = 1.5$

See Figs. B.1, B.2 and Table B.1.

B.2 Results Tables for the Parameterisation of the Darwin Genetic Algorithm

B.2.1 Icosahedron: LSF

See Tables B.2, B.3 and B.4.

B.2.2 Icosahedron: Covariance

See Tables B.5, B.6 and B.7.

B.2.3 Decahedron: LSF

See Tables B.8, B.9 and B.10.

A. J. Logsdail, *Computational Characterisation of Gold Nanocluster Structures*, Springer Theses, DOI: 10.1007/978-3-319-01493-7,

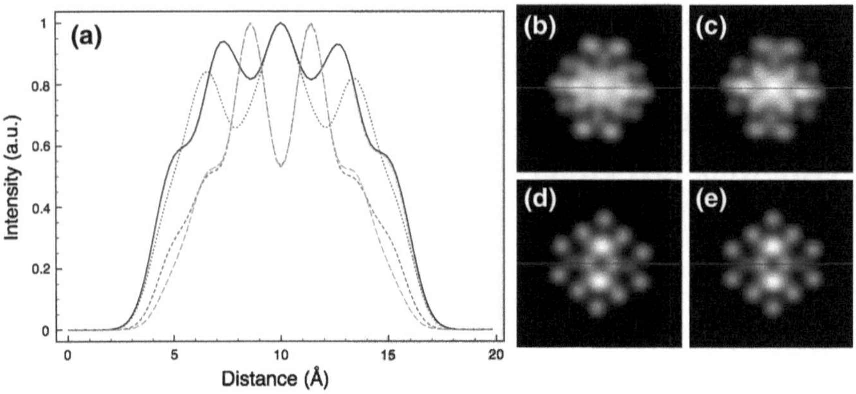

Fig. B.1 **a** Normalised linear intensity profiles for optimal orientations of Au_{38}-core(a), as given in Table B.1. Profiles given where (θ, ϕ, ψ) is: **b** $(60°, 0°, 275°)$ (*black, solid line*); **c** $(180°, -65°, 185°)$ (*red, dotted line*); **d** $(180°, -45°, 275°)$ (*blue, dashed line*); and **e** $(180°, -45°, 270°)$ (*green, long-dashed line*). Intensity profiles have been taken horizontally along the *red lines* indicated i.e. through the middle of the image

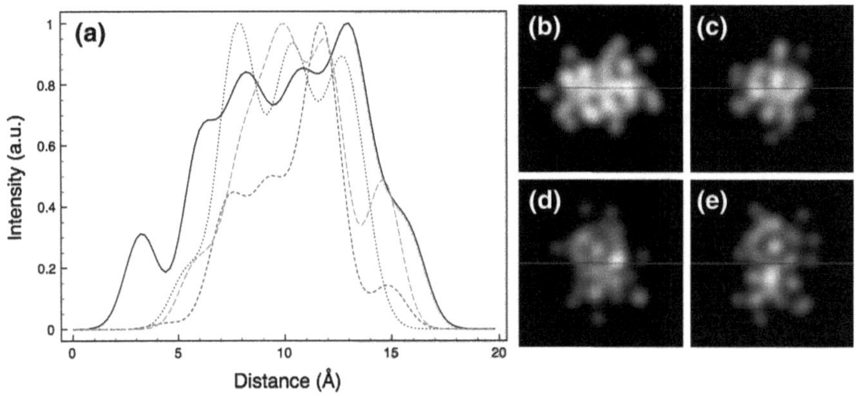

Fig. B.2 **a** Normalised linear intensity profiles for optimal orientations of Au_{38}-core(b), as given in Table B.1. Profiles given where (θ, ϕ, ψ) is: **b** $(120°, -5°, 265°)$ (*black, solid line*); **c** $(10°, -20°, 315°)$ (*red, dotted line*); **d** $(340°, 45°, 280°)$ (*blue, dashed line*); and **e** $(300°, 45°, 160°)$ (*green, long-dashed line*). Intensity profiles have been taken horizontally along the *red lines* indicated i.e. through the middle of the image

B.2.4 Decahedron: Covariance

See Tables B.11, B.12 and B.13.

Table B.1 Results, simulating Au_{38}-core structures using $\alpha = 1.5$, for the optimal orientation which minimises the chosen FE method, taken using an exhaustive search with 5° steps over the orientational space

Structure	Background threshold	FE method	Minima (θ, ϕ, ψ)	$\sigma_{\theta,\phi,\psi}$
MP-Au_{38}(a)	0	LSF	60°, 0°, 275°	953.913
	50	LSF	180°, −65°, 185°	881.419
	0	Covariance	180°, −45°, 275°	1765.72
	50	Covariance	180°, −45°, 270°	1599.74
MP-Au_{38}(b)	0	LSF	120°, −5°, 265°	827.643
	50	LSF	10°, −20°, 315°	524.514
	0	Covariance	340°, 45°, 280°	1785.93
	50	Covariance	300°, 45°, 160°	1635.57

B.3 Results Tables for the Parameterisation of the Lamarckian Genetic Algorithm

B.3.1 Icosahedron: LSF, Univariate Local Minimisation

See Tables B.14, B.15 and B.16.

B.3.2 Icosahedron: LSF, Powell's Method Local Minimisation

See Tables B.17, B.18 and B.19.

B.3.3 Icosahedron: LSF, Multivariate Local Minimisation

See Tables B.20, B.21 and B.22.

B.3.4 Icosahedron: Covariance, Multivariate Local Minimisation

See Tables B.23, B.24 and B.25.

Table B.2 Table of data from Darwinian GA calculations, using least squares fitting (LSF) for function evaluation (FE), to optimise the orientation variables θ, ψ and ϕ for an icosahedral (Ih) structure when compared to a model image

Offspring rate	0.00	0.10	0.20	0.30	0.40	0.50	0.60	0.70	0.80	0.90	1.00
Convergence	2.00	1.00	7.00	5.00	4.00	16.00	23.00	19.00	21.00	34.00	31.00
Min FE	5342.00	10847.00	9846.00	9282.00	11166.00	10512.00	6665.00	12923.00	8767.00	7873.00	9333.00
Mean FE	6118.00	10847.00	14164.57	13665.20	18327.00	16736.38	18613.74	20227.68	21134.86	23174.79	23089.35
Max FE	6894.00	10847.00	16226.00	15842.00	21577.00	21123.00	27992.00	26785.00	26486.00	31958.00	31933.00
95 % confidence	1520.96	–	1707.57	2287.35	4724.27	1708.16	2207.53	1752.36	1960.32	1555.12	1713.10
Min HE	22258.00	38453.00	27754.00	21918.00	23634.00	19588.00	10735.00	20577.00	12433.00	10127.00	11667.00
Mean HE	26282.00	38453.00	41435.43	33734.80	41272.25	33443.50	34942.43	36271.11	35682.81	39356.53	36516.26
Max HE	30306.00	38453.00	48055.00	39358.00	50023.00	43232.00	59706.00	56818.00	48817.00	65842.00	59467.00
95 % confidence	7887.04	–	5514.82	6152.60	11700.09	3903.44	4917.14	4373.53	4140.38	3886.06	3875.96
Min FEN	15.00	30.00	19.00	13.00	14.00	10.00	2.00	10.00	3.00	1.00	2.00
Mean FEN	19.00	30.00	31.86	23.80	29.50	21.81	22.09	22.11	20.81	22.21	19.55
Max FEN	23.00	30.00	38.00	29.00	37.00	30.00	41.00	36.00	30.00	39.00	34.00
95 % confidence	7.84	–	5.16	5.63	10.26	3.30	3.95	3.21	3.04	2.58	2.53

In this specific work the genetic operator under investigation is crossover. The **offspring rate** is given as a percentage function of population size, where 0.1 means for a population of 100, 10 offspring are generated. **Convergence** to the global minimum (GM) is given as a percentage (%). For those calculations which have converged to the GM, the **minimum**, **mean** and **maximum** number of **FEs** are then given, along with a **95 % confidence level**; the **minimum**, **mean** and **maximum** number of historical evaluations (**HEs**) are then given, along with a **95 % confidence level**; and finally, the **minimum**, **mean** and **maximum** number of generations completed before encountering the GM (**FEN**) are given, along with a **95 % confidence level**

Table B.3 Table of data from Darwinian GA calculations, using least squares fitting (LSF) for function evaluation (FE), to optimise the orientation variables θ, ψ and ϕ for an icosahedral (Ih) structure when compared to a model image

Mutation rate	0.00	0.10	0.20	0.30	0.40	0.50	0.60	0.70	0.80	0.90	1.00
Convergence	2.00	9.00	24.00	41.00	39.00	49.00	60.00	63.00	66.00	76.00	75.00
Min FE	6832.00	14290.00	11167.00	12250.00	9701.00	12826.00	13816.00	18589.00	17270.00	21033.00	18084.00
Mean FE	7674.00	18176.11	21796.25	23401.00	25669.36	27645.45	29025.27	29917.03	31280.55	33271.66	34285.77
Max FE	8516.00	22070.00	31122.00	33967.00	42364.00	38367.00	39772.00	40885.00	53242.00	48861.00	54103.00
95 % confidence	1650.32	1793.58	1842.71	1374.96	1926.26	1460.70	1503.44	1275.00	1788.82	1462.58	1601.71
Min HE	12968.00	24810.00	16033.00	16350.00	11499.00	13974.00	14184.00	18043.00	15730.00	18475.00	14707.00
Mean HE	14819.50	35912.56	37234.63	35008.68	35130.00	34356.98	33533.48	31703.00	31461.67	31756.88	31030.57
Max HE	16671.00	53423.00	62039.00	56565.00	66836.00	50533.00	53114.00	46480.00	60358.00	52905.00	62697.00
95 % confidence	3628.94	6244.09	4293.89	2756.60	3355.40	2289.13	2237.04	1671.01	2316.58	1793.85	1870.34
Min FEN	3.00	13.00	6.00	6.00	2.00	4.00	4.00	7.00	5.00	7.00	4.00
Mean FEN	4.50	20.89	21.92	20.20	20.00	19.31	18.40	16.97	16.44	16.38	15.61
Max FEN	6.00	32.00	39.00	35.00	42.00	31.00	31.00	27.00	36.00	30.00	34.00
95 % confidence	2.94	4.20	3.06	1.96	2.40	1.63	1.55	1.18	1.57	1.20	1.24

In this specific work the genetic operator under investigation is mutation of parents. The **mutation rate** is given as a percentage function of population size, where 0.1 means for a population of 100, 10 mutants are generated. **Convergence** to the global minimum (GM) is given as a percentage (%). For those calculations which have converged to the GM, the **minimum**, **mean** and **maximum** number of **FEs** are then given, along with a **95 % confidence level**; the **minimum**, **mean** and **maximum** number of historical evaluations (**HEs**) are then given, along with a **95 % confidence level**; and finally, the **minimum**, **mean** and **maximum** number of generations completed before encountering the GM (**FEN**) are given, along with a **95 % confidence level**

Table B.4 Table of data from Darwinian GA calculations, using least squares fitting (LSF) for function evaluation (FE), to optimise the orientation variables θ, ψ and ϕ for an icosahedral (Ih) structure when compared to a model image. In this specific work the genetic operator under investigation is mutation of offspring

Mutation rate	0.00	0.10	0.20	0.30	0.40	0.50	0.60	0.70	0.80	0.90	1.00
Convergence 6.00	9.00	15.00	19.00	17.00	18.00	28.00	32.00	33.00	28.00	32.00	
Min FE	10178.00	12390.00	8041.00	18308.00	13604.00	12721.00	12721.00	13687.00	18536.00	22721.00	22783.00
Mean FE	15211.83	18791.33	18744.33	24295.00	23459.24	26268.17	30139.29	29127.66	33377.36	33275.36	40065.84
Max FE	18110.00	21430.00	23717.00	28114.00	29816.00	35741.00	41162.00	38881.00	46320.00	54030.00	56094.00
95 % confidence	2457.24	1946.60	2217.45	1419.52	2239.95	2403.54	2300.57	2415.40	2555.36	2409.87	2944.73
Min HE	20422.00	21010.00	11159.00	24992.00	16386.00	14079.00	12879.00	12889.00	16835.00	19603.00	18417.00
Mean HE	38488.17	37197.56	31255.20	37020.42	30870.18	31581.50	34632.04	30384.47	32871.12	30338.07	35520.50
Max HE	47951.00	44279.00	47554.00	49710.00	41984.00	46259.00	53517.00	47819.00	51680.00	58570.00	57906.00
95 % confidence	9434.99	5011.29	4833.54	3190.58	3439.55	3465.65	3452.13	2961.15	3128.86	2880.61	3317.28
Min FEN	9.00	10.00	2.00	13.00	6.00	4.00	3.00	3.00	6.00	8.00	7.00
Mean FEN	21.83	21.89	17.40	21.58	17.06	17.50	19.32	16.13	17.79	15.86	19.28
Max FEN	28.00	27.00	28.00	29.00	25.00	28.00	31.00	27.00	30.00	34.00	33.00
95 % confidence	6.59	3.66	3.51	2.18	2.58	2.55	2.39	2.14	2.18	1.95	2.23

The **mutation rate** is given as a percentage function of population size, where 0.1 means for a population of 100, 10 mutants are generated. **Convergence** to the global minimum (GM) is given as a percentage (%). For those calculations which have converged to the GM, the **minimum**, **mean** and **maximum** number of **FEs** are then given, along with a **95 % confidence level**; the **minimum**, **mean** and **maximum** number of historical evaluations (**HEs**) are then given, along with a **95 % confidence level**; and finally, the **minimum**, **mean** and **maximum** number of generations completed before encountering the GM (**FEN**) are given, along with a **95 % confidence level**

Table B.5 Table of data from Darwinian GA calculations, using covariance for function evaluation (FE), to optimise the orientation variables θ, ψ and ϕ for an icosahedral (Ih) structure when compared to a model image

Offspring rate	0.00	0.10	0.20	0.30	0.40	0.50	0.60	0.70	0.80	0.90	1.00
Convergence	1.00	3.00	3.00	8.00	8.00	11.00	23.00	32.00	36.00	41.00	44.00
Min FE	5345.00	4253.00	7216.00	8896.00	8297.00	8766.00	9709.00	10786.00	7240.00	7785.00	9271.00
Mean FE	5345.00	6392.67	14246.67	13090.13	17753.50	21423.27	19701.39	20918.81	19870.94	21811.27	22806.11
Max FE	5345.00	10376.00	18744.00	19137.00	22626.00	29046.00	27254.00	28039.00	29625.00	30461.00	29905.00
95 % confidence	–	3907.26	6979.09	2437.85	3112.11	3887.33	2017.62	1385.54	1826.13	1293.43	1205.81
Min HE	22255.00	12547.00	19184.00	20804.00	16903.00	16234.00	16691.00	17014.00	9960.00	10215.00	11729.00
Mean HE	22255.00	20807.33	42017.33	32540.75	41246.50	46849.45	39097.17	39120.38	34773.00	37983.61	39143.25
Max HE	22255.00	36324.00	56648.00	51017.00	55374.00	67376.00	56746.00	56656.00	59575.00	63139.00	61495.00
95 % confidence	–	15216.94	22670.31	7114.53	8317.13	9862.70	5007.60	3467.97	4096.54	3021.39	3014.16
Min FEN	15.00	5.00	11.00	12.00	8.00	7.00	7.00	7.00	1.00	1.00	2.00
Mean FEN	15.00	13.00	32.33	22.63	29.13	32.45	25.00	23.97	19.72	20.90	20.61
Max FEN	15.00	28.00	46.00	39.00	41.00	49.00	39.00	36.00	37.00	37.00	34.00
95 % confidence	–	14.71	21.18	6.37	7.14	8.08	3.89	2.55	2.96	2.04	1.90

In this specific work the genetic operator under investigation is mating. Table contents are as defined for Table B.2

Table B.6 Table of data from Darwinian GA calculations, using covariance for function evaluation (FE), to optimise the orientation variables θ, ψ and ϕ for an icosahedral (Ih) structure when compared to a model image

Mutation rate	0.00	0.10	0.20	0.30	0.40	0.50	0.60	0.70	0.80	0.90	1.00
Convergence	8.00	22.00	35.00	47.00	56.00	59.00	68.00	77.00	84.00	87.00	87.00
Min FE	13634.00	10411.00	8824.00	12234.00	11714.00	11630.00	14863.00	13605.00	17192.00	15503.00	13381.00
Mean FE	15396.38	19937.45	20336.17	23328.85	25915.04	27121.42	28249.96	28658.90	29673.04	31931.22	32439.33
Max FE	17941.00	24997.00	29124.00	31830.00	38720.00	39625.00	40509.00	47579.00	40173.00	43616.00	46956.00
95 % confidence	953.34	1474.96	1836.02	1330.66	1344.10	1459.27	1230.22	1319.35	1033.74	1058.54	1078.39
Min HE	31366.00	17289.00	12376.00	16366.00	13886.00	12870.00	15537.00	13095.00	15808.00	13397.00	11019.00
Mean HE	41528.63	44466.00	36005.83	36636.53	36926.14	35074.42	33349.96	31902.51	30688.06	31664.82	30516.33
Max HE	59459.00	65403.00	60076.00	54879.00	68280.00	61433.00	57091.00	66596.00	47427.00	52032.00	53013.00
95 % confidence	5999.83	4997.43	4313.83	2721.74	2600.98	2434.62	1938.40	1979.09	1376.96	1493.87	1406.04
Min FEN	17.00	7.00	3.00	6.00	4.00	3.00	5.00	3.00	5.00	3.00	1.00
Mean FEN	23.63	26.32	20.57	20.94	20.93	19.39	18.00	16.55	15.52	15.85	14.77
Max FEN	35.00	40.00	37.00	33.00	41.00	36.00	33.00	38.00	26.00	27.00	28.00
95 % confidence	3.85	3.39	3.06	1.92	1.78	1.69	1.32	1.31	0.92	0.94	0.88

In this specific work the genetic operator under investigation is mutation of parents. Table contents are as defined for Table B.3

Table B.7 Table of data from Darwinian GA calculations, using covariance for function evaluation (FE), to optimise the orientation variables θ, ψ and ϕ for an icosahedral (Ih) structure when compared to a model image

Mutation rate	0.00	0.10	0.20	0.30	0.40	0.50	0.60	0.70	0.80	0.90	1.00
Convergence	4.00	15.00	18.00	31.00	37.00	37.00	48.00	41.00	44.00	47.00	49.00
Min FE	10958.00	10971.00	10256.00	11452.00	11683.00	22540.00	13813.00	13591.00	18578.00	14086.00	16573.00
Mean FE	14893.50	18285.07	21635.50	24117.45	26275.41	30483.49	32289.42	33375.32	35689.14	35321.87	39332.57
Max FE	17377.00	23608.00	28369.00	33043.00	35551.00	45821.00	43896.00	43354.00	48425.00	53347.00	55347.00
95 % confidence	2919.94	1593.72	2086.27	1719.89	1818.40	1765.11	1831.88	2002.04	1871.32	2180.96	2386.73
Min HE	23242.00	18629.00	14944.00	15048.00	13917.00	26923.00	14187.00	13109.00	17022.00	12114.00	13427.00
Mean HE	44505.75	40068.27	42453.39	40181.45	38744.68	40822.24	39207.58	37408.49	36722.77	33904.85	35752.08
Max HE	62180.00	59507.00	74831.00	66957.00	58059.00	66079.00	58504.00	51983.00	54775.00	59102.00	55853.00
95 % confidence	19165.77	5586.26	6369.95	4271.53	3492.28	3172.09	2798.77	2766.52	2392.51	2637.99	2655.11
Min FEN	11.00	8.00	5.00	5.00	4.00	14.00	4.00	3.00	6.00	2.00	3.00
Mean FEN	25.00	23.13	24.44	23.00	21.92	23.35	22.13	20.63	20.16	17.94	19.10
Max FEN	36.00	36.00	44.00	40.00	34.00	41.00	35.00	30.00	32.00	33.00	32.00
95 % confidence	12.21	3.74	4.19	2.83	2.40	2.13	1.92	1.90	1.63	1.78	1.80

In this specific work the genetic operator under investigation is mutation of offspring. Table contents are as defined for Table B.4

Table B.8 Table of data from Darwinian GA calculations, using least squares fitting (LSF) for function evaluation (FE), to optimise the orientation variables θ, ψ and ϕ for a decahedral (Dh) structure when compared to a model image

Offspring rate	0.00	0.10	0.20	0.30	0.40	0.50	0.60	0.70	0.80	0.90	1.00
Convergence	1.00	1.00	3.00	11.00	26.00	30.00	43.00	60.00	64.00	72.00	79.00
Min FE	3762.00	5336.00	6130.00	10077.00	8839.00	5573.00	9169.00	8059.00	10965.00	11023.00	13272.00
Mean FE	3762.00	5336.00	11715.00	17052.91	15757.19	17591.37	18427.60	18179.42	18834.63	18362.07	19148.41
Max FE	3762.00	5336.00	16556.00	19841.00	21226.00	22502.00	23270.00	22324.00	22389.00	22840.00	23137.00
95 % confidence	–	–	5943.95	1639.09	1362.79	1407.31	832.51	681.01	464.08	496.24	358.80
Min HE	14238.00	16664.00	16070.00	24123.00	17961.00	9227.00	15431.00	12141.00	16235.00	15362.00	18728.00
Mean HE	14238.00	16664.00	34751.67	49737.00	41873.46	50017.70	50082.79	48789.25	50426.05	47003.82	50946.70
Max HE	14238.00	16664.00	51844.00	61568.00	64013.00	78172.00	80838.00	73166.00	68938.00	74686.00	78980.00
95 % confidence	–	–	20300.88	6061.98	5435.75	6639.45	4598.48	3596.62	3039.09	2953.10	2679.79
Min FEN	7.00	9.00	8.00	15.00	9.00	1.00	6.00	3.00	6.00	5.00	7.00
Mean FEN	7.00	9.00	25.33	36.73	28.27	32.07	30.40	27.62	27.03	23.56	24.32
Max FEN	7.00	9.00	41.00	46.00	45.00	51.00	49.00	42.00	37.00	38.00	38.00
95 % confidence	–	–	18.74	5.11	4.24	4.70	2.99	2.22	1.72	1.62	1.36

In this specific work the genetic operator under investigation is mating. Table contents are as defined for Table B.2

Table B.9 Table of data from Darwinian GA calculations, using least squares fitting (LSF) for function evaluation (FE), to optimise the orientation variables θ, ψ and ϕ for a decahedral (Dh) structure when compared to a model image

Mutation rate	0.00	0.10	0.20	0.30	0.40	0.50	0.60	0.70	0.80	0.90	1.00
Convergence	27.00	53.00	60.00	67.00	78.00	80.00	81.00	90.00	87.00	92.00	91.00
Min FE	9749.00	10739.00	11454.00	11285.00	12471.00	13615.00	10265.00	17071.00	18063.00	13998.00	23328.00
Mean FE	12292.52	15679.32	18136.37	20538.64	23422.23	25305.50	27497.78	29996.29	31594.80	32088.02	41797.09
Max FE	13651.00	19226.00	22130.00	25024.00	29563.00	31388.00	35878.00	39994.00	41156.00	41803.00	98385.00
95 % confidence	322.46	453.25	580.90	671.45	736.25	702.33	847.66	832.28	840.66	1002.66	2955.85
Min HE	20851.00	18861.00	17493.00	15215.00	15329.00	15485.00	10535.00	17129.00	17537.00	12202.00	20672.00
Mean HE	46039.59	47941.23	46828.63	46393.67	46431.18	45854.44	45863.94	47396.77	46620.70	43298.24	61055.84
Max HE	63219.00	82008.00	83029.00	79176.00	110437.00	72267.00	84816.00	77180.00	93932.00	84297.00	208065.00
95 % confidence	3972.36	3696.76	3638.63	3160.16	3236.68	2617.45	2663.97	2557.43	2752.94	2584.63	7177.63
Min FEN	9.00	8.00	7.00	5.00	5.00	5.00	1.00	6.00	6.00	2.00	1.00
Mean FEN	24.41	25.91	24.88	24.25	24.12	23.29	22.90	23.28	22.39	20.22	19.30
Max FEN	34.00	45.00	45.00	42.00	56.00	36.00	42.00	38.00	44.00	39.00	36.00
95 % confidence	2.29	2.16	2.08	1.80	1.77	1.43	1.44	1.34	1.36	1.31	1.60

In this specific work the genetic operator under investigation is mutation of parents. Table contents are as defined for Table B.3

Table B.10 Table of data from Darwinian GA calculations, using least squares fitting (LSF) for function evaluation (FE), to optimise the orientation variables θ, ψ and ϕ for a decahedral (Dh) structure when compared to a model image

Mutation rate	0.00	0.10	0.20	0.30	0.40	0.50	0.60	0.70	0.80	0.90	1.00
Convergence	29.00	46.00	59.00	57.00	63.00	66.00	70.00	71.00	71.00	78.00	75.00
Min FE	5689.00	9627.00	14388.00	13913.00	15681.00	11459.00	22511.00	11040.00	20355.00	19243.00	17966.00
Mean FE	12164.17	15712.83	18286.02	21367.19	23309.65	27286.88	29939.39	30195.63	34292.54	34391.88	37593.72
Max FE	13859.00	19390.00	22081.00	25640.00	30147.00	34920.00	37135.00	39019.00	47671.00	48119.00	55502.00
95 % confidence	534.86	513.98	444.51	663.57	689.15	888.25	784.41	1252.33	1158.89	1373.48	1455.50
Min HE	10511.00	16173.00	24812.00	18887.00	20919.00	13041.00	26559.00	10660.00	20445.00	17200.00	14834.00
Mean HE	45621.62	47547.72	47218.05	48567.96	45765.83	50356.32	51235.67	44848.17	48509.62	44110.65	45455.73
Max HE	70741.00	82410.00	78144.00	74360.00	72453.00	83641.00	83149.00	66232.00	84106.00	77241.00	83698.00
95 % confidence	4408.30	3950.43	3171.80	3217.25	2936.68	2968.08	2807.39	2974.34	2850.42	2986.99	3062.52
Min FEN	1.00	6.00	12.00	8.00	9.00	3.00	13.00	1.00	8.00	6.00	4.00
Mean FEN	24.10	25.72	25.15	25.68	23.76	26.11	26.16	22.34	24.15	21.37	21.95
Max FEN	39.00	46.00	42.00	40.00	39.00	43.00	42.00	34.00	43.00	37.00	42.00
95 % confidence	2.67	2.31	1.78	1.83	1.62	1.65	1.47	1.67	1.53	1.60	1.60

In this specific work the genetic operator under investigation is mutation of offspring. Table contents are as defined for Table B.4

Table B.11 Table of data from Darwinian GA calculations, using covariance for function evaluation (FE), to optimise the orientation variables θ, ψ and ϕ for a decahedral (Dh) structure when compared to a model image

Offspring rate	0.00	0.10	0.20	0.30	0.40	0.50	0.60	0.70	0.80	0.90	1.00
Convergence	1.00	1.00	2.00	6.00	16.00	23.00	30.00	43.00	55.00	59.00	63.00
Min FE	2582.00	21239.00	14653.00	16680.00	11711.00	8830.00	7332.00	12887.00	15801.00	14980.00	15234.00
Mean FE	2582.00	21239.00	16523.00	17880.17	17758.00	16556.48	17727.20	17345.51	18024.78	17686.41	17963.57
Max FE	2582.00	21239.00	18393.00	19215.00	21494.00	19961.00	20054.00	20889.00	20142.00	20489.00	21085.00
95 % confidence	–	–	3665.20	909.21	1189.34	1217.27	818.92	514.89	291.52	292.32	270.71
Min HE	8218.00	89161.00	46747.00	46020.00	26289.00	16170.00	11868.00	22513.00	33399.00	26120.00	25384.00
Mean HE	8218.00	89161.00	54677.00	54569.83	53341.25	45768.35	48811.80	45361.47	47282.11	45658.22	46023.49
Max HE	8218.00	89161.00	62607.00	59996.00	75706.00	62307.00	63946.00	61005.00	63296.00	61473.00	67583.00
95 % confidence	–	–	15542.80	4184.24	6565.64	5450.16	3818.13	2774.85	1907.49	1851.17	1907.75
Min FEN	1.00	77.00	36.00	34.00	16.00	7.00	3.00	11.00	17.00	12.00	11.00
Mean FEN	1.00	77.00	43.00	40.50	36.69	28.96	29.30	25.37	25.05	22.59	21.54
Max FEN	1.00	77.00	50.00	45.00	53.00	40.00	39.00	35.00	34.00	31.00	31.00
95 % confidence	–	–	13.72	3.35	4.82	3.90	2.54	1.71	1.08	1.00	0.95

In this specific work the genetic operator under investigation is mating. Table contents are as defined for Table B.2

Table B.12 Table of data from Darwinian GA calculations, using covariance for function evaluation (FE), to optimise the orientation variables θ, ψ and ϕ for a decahedral (Dh) structure when compared to a model image

Mutation rate	0.00	0.10	0.20	0.30	0.40	0.50	0.60	0.70	0.80	0.90	1.00
Convergence	14.00	31.00	53.00	58.00	71.00	69.00	72.00	82.00	84.00	92.00	87.00
Min FE	10334.00	13549.00	14708.00	15265.00	14075.00	19344.00	17589.00	17857.00	23025.00	12619.00	16402.00
Mean FE	11567.64	15376.65	17902.42	19946.47	21891.77	23798.99	24983.39	26241.62	28083.81	28858.90	30504.38
Max FE	12272.00	17216.00	20512.00	23282.00	25896.00	29359.00	29393.00	32046.00	34112.00	35910.00	36434.00
95 % confidence	319.55	312.04	351.94	440.76	491.88	467.63	466.39	503.11	534.69	656.88	636.24
Min HE	23866.00	32836.00	31670.00	28035.00	18125.00	27682.00	20011.00	18843.00	25575.00	10881.00	13598.00
Mean HE	47573.07	50567.03	47522.34	45584.45	43927.56	43100.14	40915.68	40091.78	40324.46	39396.07	39853.78
Max HE	63569.00	64448.00	69988.00	64264.00	58250.00	57493.00	56227.00	52154.00	57389.00	57790.00	54686.00
95 % confidence	5484.45	2935.54	2190.63	2220.09	1949.86	1628.77	1492.38	1452.71	1490.12	1626.74	1532.69
Min FEN	11.00	17.00	16.00	13.00	7.00	13.00	8.00	7.00	11.00	1.00	3.00
Mean FEN	24.86	27.13	25.11	23.59	22.28	21.43	19.79	18.85	18.62	17.58	17.41
Max FEN	34.00	35.00	37.00	34.00	30.00	29.00	27.00	26.00	27.00	27.00	24.00
95 % confidence	3.10	1.67	1.24	1.24	1.09	0.89	0.80	0.76	0.76	0.83	0.76

In this specific work the genetic operator under investigation is mutation of parents. Table contents are as defined for Table B.3

Table B.13 Table of data from Darwinian GA calculations, using covariance for function evaluation (FE), to optimise the orientation variables θ, ψ and ϕ for a decahedral (Dh) structure when compared to a model image

Mutation rate	0.00	0.10	0.20	0.30	0.40	0.50	0.60	0.70	0.80	0.90	1.00
Convergence	14.00	28.00	26.00	49.00	49.00	56.00	63.00	65.00	58.00	59.00	68.00
Min FE	10580.00	11703.00	8734.00	16955.00	17938.00	19848.00	21109.00	23283.00	19201.00	23353.00	28742.00
Mean FE	11709.93	14725.89	17975.38	20531.88	22883.82	25283.20	27160.41	29522.63	30841.50	33411.46	34563.19
Max FE	12802.00	16726.00	21005.00	23240.00	28981.00	31468.00	32803.00	35579.00	37708.00	40517.00	41622.00
95 % confidence	374.26	484.17	992.91	447.42	588.56	595.23	659.61	623.04	975.66	843.18	667.21
Min HE	28185.00	21697.00	12466.00	28194.00	27242.00	27652.00	26091.00	28417.00	18999.00	21747.00	28700.00
Mean HE	46660.79	47241.25	49375.54	46638.73	46310.39	44928.27	43733.57	45023.25	42770.38	43218.39	42418.57
Max HE	63454.00	73765.00	64728.00	66990.00	64819.00	65688.00	60768.00	62265.00	61067.00	63752.00	62429.00
95 % confidence	5511.93	4730.02	4936.51	2530.20	2202.02	2011.41	2040.98	1840.23	2356.61	1921.98	1610.35
Min FEN	14.00	10.00	3.00	8.00	13.00	13.00	12.00	13.00	7.00	9.00	13.00
Mean FEN	24.43	25.04	26.08	23.88	23.82	22.88	21.87	22.14	20.62	20.68	19.78
Max FEN	34.00	40.00	35.00	35.00	35.00	34.00	31.00	31.00	29.00	28.00	28.00
95 % confidence	3.13	2.73	2.92	1.50	1.25	1.10	1.11	0.96	1.26	0.99	0.78

In this specific work the genetic operator under investigation is mutation of offspring. Table contents are as defined for Table B.4

Table B.14 Table of data from Lamarckian GA calculations, using least squares fitting (LSF) for function evaluation (FE) and a univariate local minimisation routine, to optimise the orientation variables θ, ψ and ϕ for an icosahedral (Ih) structure when compared to a model image

Offspring rate	0.00	0.10	0.20	0.30	0.40	0.50	0.60	0.70	0.80	0.90	1.00
Convergence 73.00	80.00	82.00	81.00	82.00	90.00	84.00	85.00	87.00	88.00	88.00	
Min FE	1987.00	2375.00	2636.00	2933.00	3242.00	3397.00	3779.00	3920.00	3946.00	3976.00	3821.00
Mean FE	2465.32	3243.60	3599.45	3984.32	4295.54	4685.70	4986.60	5169.07	5453.78	5727.91	5940.55
Max FE	3951.00	6040.00	6309.00	5604.00	6811.00	6700.00	7663.00	7833.00	8225.00	7935.00	8066.00
95 % confidence	117.01	165.28	152.61	126.55	140.17	146.20	150.77	147.88	184.55	157.38	172.41
Min HE	5575.00	6208.00	6895.00	7660.00	8115.00	8570.00	9550.00	10027.00	10356.00	11279.00	11491.00
Mean HE	7147.01	8793.30	9047.63	9716.16	10404.90	11403.73	12174.23	12729.81	13505.60	14265.59	15407.22
Max HE	14156.00	18821.00	17410.00	14331.00	19130.00	18311.00	21623.00	23526.00	24028.00	23862.00	23794.00
95 % confidence	455.70	582.33	479.41	373.50	428.40	452.20	473.76	491.93	548.79	466.99	547.89
Min FEN	1.00	1.00	1.00	1.00	1.00	1.00	1.00	1.00	1.00	1.00	1.00
Mean FEN	3.05	4.03	3.04	2.77	2.63	2.74	2.61	2.38	2.30	2.30	2.53
Max FEN	14.00	17.00	14.00	8.00	11.00	9.00	10.00	11.00	9.00	9.00	8.00
95 % confidence	0.74	0.85	0.59	0.43	0.43	0.43	0.42	0.39	0.37	0.34	0.36

In this specific work the genetic operator under investigation is mating. Table contents are as defined for Table B.2

Table B.15 Table of data from Lamarckian GA calculations, using least squares fitting (LSF) for function evaluation (FE) and a univariate local minimisation routine, to optimise the orientation variables θ, ψ and ϕ for an icosahedral (Ih) structure when compared to a model image

Mutation rate	0.00	0.10	0.20	0.30	0.40	0.50	0.60	0.70	0.80	0.90	1.00
Convergence	83.00	85.00	80.00	85.00	89.00	91.00	91.00	85.00	94.00	93.00	92.00
Min FE	3149.00	3300.00	3714.00	4839.00	5026.00	5192.00	5946.00	6265.00	6449.00	6695.00	7071.00
Mean FE	4046.73	4585.72	5495.36	6101.19	6645.45	7168.58	7622.81	7777.99	8403.21	8576.60	8877.41
Max FE	5530.00	6548.00	8560.00	8398.00	9549.00	11485.00	12061.00	10348.00	12675.00	11241.00	13812.00
95 % confidence	110.39	120.54	211.80	177.95	197.12	181.24	206.48	201.15	249.40	215.18	231.32
Min HE	8878.00	9351.00	10218.00	11976.00	12815.00	13429.00	14607.00	15681.00	16564.00	17478.00	18421.00
Mean HE	11005.86	11971.11	14081.14	14987.72	16029.16	17104.85	18379.01	19142.21	20789.15	21578.53	22498.02
Max HE	17775.00	18493.00	27821.00	24524.00	26559.00	32354.00	28400.00	31025.00	32577.00	30344.00	36207.00
95 % confidence	341.05	409.87	697.12	555.29	596.29	536.23	585.28	603.78	690.06	597.12	584.50
Min FEN	1.00	1.00	1.00	1.00	1.00	1.00	1.00	1.00	1.00	1.00	1.00
Mean FEN	2.29	2.28	2.84	2.52	2.38	2.27	2.29	2.02	2.32	2.22	2.04
Max FEN	9.00	9.00	13.00	8.00	9.00	10.00	7.00	7.00	7.00	6.00	7.00
95 % confidence	0.34	0.35	0.52	0.35	0.35	0.30	0.31	0.27	0.29	0.23	0.22

In this specific work the genetic operator under investigation is mutation of parents. Table contents are as defined for Table B.3

Table B.16 Table of data from Lamarckian GA calculations, using least squares fitting (LSF) for function evaluation (FE) and a univariate local minimisation routine, to optimise the orientation variables θ, ψ and ϕ for an icosahedral (Ih) structure when compared to a model image

Mutation rate	0.00	0.10	0.20	0.30	0.40	0.50	0.60	0.70	0.80	0.90	1.00
Convergence	80.00	81.00	84.00	89.00	88.00	94.00	92.00	92.00	95.00	97.00	92.00
Min FE	3201.00	3473.00	3970.00	4254.00	4669.00	5009.00	6124.00	6175.00	7032.00	7447.00	7400.00
Mean FE	3965.50	4695.79	5375.39	5900.16	6652.15	7501.03	8077.70	8630.05	9430.43	9747.47	10276.83
Max FE	5086.00	6803.00	6788.00	8430.00	10185.00	11149.00	12289.00	12710.00	13945.00	16778.00	13488.00
95 % confidence	95.36	151.99	135.32	151.79	194.36	253.81	228.16	255.09	281.73	294.07	276.76
Min HE	9121.00	9203.00	10506.00	11247.00	12456.00	12942.00	15174.00	15610.00	17059.00	17881.00	19213.00
Mean HE	10954.66	12630.99	13918.29	14826.44	16257.59	17944.36	19085.12	20066.77	21991.09	22609.32	24338.35
Max HE	17402.00	19082.00	21660.00	23875.00	28899.00	29161.00	29571.00	31420.00	34093.00	43156.00	34523.00
95 % confidence	341.05	480.29	512.76	505.21	576.72	754.35	654.42	632.85	699.67	728.44	711.82
Min FEN	1.00	1.00	1.00	1.00	1.00	1.00	1.00	1.00	1.00	1.00	1.00
Mean FEN	2.31	2.79	2.93	2.48	2.58	2.65	2.43	2.24	2.48	2.16	2.26
Max FEN	8.00	9.00	10.00	9.00	9.00	9.00	9.00	8.00	7.00	10.00	7.00
95 % confidence	0.37	0.42	0.44	0.35	0.36	0.41	0.36	0.28	0.32	0.28	0.27

In this specific work the genetic operator under investigation is mutation of offspring. Table contents are as defined for Table B.4

Table B.17 Table of data from Lamarckian GA calculations, using least squares fitting (LSF) for function evaluation (FE) and a Powell's local minimisation routine, to optimise the orientation variables θ, ψ and ϕ for an icosahedral (Ih) structure when compared to a model image

Offspring rate	0.00	0.10	0.20	0.30	0.40	0.50	0.60	0.70	0.80	0.90	1.00
Convergence	84.00	78.00	84.00	89.00	92.00	98.00	91.00	91.00	95.00	92.00	93.00
Min FE	2045.00	2656.00	2957.00	3302.00	3517.00	3501.00	3681.00	3808.00	4511.00	4344.00	4256.00
Mean FE	2621.18	3269.09	3729.94	4098.57	4344.60	4712.62	4862.66	5213.40	5640.69	5768.76	5835.24
Max FE	4040.00	5281.00	5627.00	5978.00	6237.00	7885.00	6333.00	7065.00	8622.00	8148.00	9183.00
95 % confidence	89.67	115.74	124.84	115.04	103.47	133.58	104.93	135.12	169.47	156.80	155.90
Min HE	6837.00	7804.00	8423.00	9045.00	10033.00	10666.00	11220.00	11843.00	12433.00	13086.00	13330.00
Mean HE	8379.32	9498.60	10329.99	11061.37	11754.58	12762.07	13182.57	14228.68	15607.07	16221.53	16640.12
Max HE	14497.00	16826.00	17594.00	17935.00	19371.00	27386.00	18888.00	22931.00	27538.00	24000.00	29364.00
95 % confidence	394.02	451.26	445.24	412.95	378.72	495.11	336.19	438.30	579.65	466.69	524.94
Min FEN	1.00	1.00	1.00	1.00	1.00	1.00	1.00	1.00	1.00	1.00	1.00
Mean FEN	2.37	2.49	2.29	2.09	1.92	2.07	1.73	1.89	2.11	1.96	1.77
Max FEN	10.00	11.00	10.00	8.00	8.00	14.00	7.00	8.00	12.00	7.00	8.00
95 % confidence	0.51	0.53	0.47	0.38	0.31	0.40	0.25	0.29	0.37	0.29	0.28

In this specific work the genetic operator under investigation is mating. Table contents are as defined for Table B.2

Table B.18 Table of data from Lamarckian GA calculations, using least squares fitting (LSF) for function evaluation (FE) and a Powell's local minimisation routine, to optimise the orientation variables θ, ψ and ϕ for an icosahedral (Ih) structure when compared to a model image

Mutation rate	0.00	0.10	0.20	0.30	0.40	0.50	0.60	0.70	0.80	0.90	1.00
Convergence	88.00	93.00	94.00	95.00	93.00	90.00	97.00	96.00	96.00	92.00	95.00
Min FE	3119.00	3802.00	4283.00	5024.00	5481.00	5744.00	5292.00	5995.00	6403.00	6559.00	7377.00
Mean FE	4263.17	4886.10	5861.03	6607.42	7010.09	7596.79	7705.90	8031.52	8444.01	8744.03	9008.26
Max FE	5628.00	7404.00	9184.00	8393.00	9907.00	11646.00	11285.00	10948.00	12672.00	12204.00	13772.00
95 % confidence	94.24	134.88	163.22	137.36	144.38	219.63	178.80	180.56	238.84	193.59	241.81
Min HE	10906.00	11517.00	12678.00	14023.00	15132.00	16195.00	17483.00	17255.00	19187.00	20166.00	21249.00
Mean HE	13038.48	14290.35	15852.33	17291.31	18329.49	20437.98	20545.51	21731.95	23292.11	24125.53	25068.20
Max HE	17922.00	27375.00	29651.00	28892.00	27599.00	47385.00	32581.00	30171.00	34743.00	32478.00	43516.00
95 % confidence	281.63	543.59	586.27	542.31	429.05	901.34	502.68	500.01	649.25	507.24	715.57
Min FEN	1.00	1.00	1.00	1.00	1.00	1.00	1.00	1.00	1.00	1.00	1.00
Mean FEN	2.01	2.12	2.05	2.03	1.88	2.30	1.76	1.73	1.86	1.72	1.68
Max FEN	7.00	11.00	10.00	9.00	7.00	15.00	7.00	5.00	6.00	4.00	8.00
95 % confidence	0.25	0.37	0.34	0.33	0.22	0.41	0.20	0.18	0.22	0.16	0.22

In this specific work the genetic operator under investigation is mutation of parents. Table contents are as defined for Table B.3

Table B.19 Table of data from Lamarckian GA calculations, using least squares fitting (LSF) for function evaluation (FE) and a Powell's local minimisation routine, to optimise the orientation variables θ, ψ and ϕ for an icosahedral (Ih) structure when compared to a model image

Mutation rate	0.00	0.10	0.20	0.30	0.40	0.50	0.60	0.70	0.80	0.90	1.00
Convergence	87.00	95.00	90.00	95.00	93.00	95.00	94.00	96.00	97.00	96.00	100.00
Min FE	3110.00	3625.00	4072.00	4875.00	5697.00	6076.00	5000.00	6874.00	7240.00	7647.00	8678.00
Mean FE	4150.15	4877.69	5597.05	6395.96	7307.68	7774.52	8478.66	9301.07	9787.88	10470.97	10905.84
Max FE	5325.00	6356.00	7760.00	9775.00	10080.00	10883.00	12422.00	12819.00	15038.00	14095.00	14391.00
95 % confidence	93.89	103.94	132.99	174.88	184.56	189.69	246.33	234.61	235.06	255.67	247.64
Min HE	10654.00	11174.00	12932.00	14074.00	15338.00	15838.00	16651.00	18497.00	20337.00	21034.00	23067.00
Mean HE	13142.26	14043.97	15256.41	16974.64	18644.00	19656.47	21386.15	23125.99	24179.95	25799.50	27276.72
Max HE	20883.00	19822.00	19178.00	28765.00	27022.00	29197.00	38881.00	37341.00	36847.00	39246.00	38079.00
95 % confidence	369.90	326.37	345.86	500.50	496.60	508.45	794.22	716.45	653.37	702.72	593.97
Min FEN	1.00	1.00	1.00	1.00	1.00	1.00	1.00	1.00	1.00	1.00	1.00
Mean FEN	2.23	1.85	1.79	1.99	1.96	1.81	1.91	1.98	1.85	1.93	1.86
Max FEN	10.00	6.00	5.00	9.00	7.00	6.00	9.00	8.00	7.00	7.00	5.00
95 % confidence	0.32	0.21	0.22	0.29	0.24	0.23	0.33	0.29	0.24	0.24	0.19

In this specific work the genetic operator under investigation is mutation of offspring. Table contents are as defined for Table B.4

Table B.20 Table of data from Lamarckian GA calculations, using least squares fitting (LSF) for function evaluation (FE) and a multivariate local minimisation routine, to optimise the orientation variables θ, ψ and ϕ for an icosahedral (Ih) structure when compared to a model image

Offspring rate	0.00	0.10	0.20	0.30	0.40	0.50	0.60	0.70	0.80	0.90	1.00
Convergence	71.00	79.00	85.00	76.00	75.00	88.00	86.00	79.00	86.00	80.00	88.00
Min FE	12141.00	13538.00	15518.00	17238.00	19444.00	20529.00	21299.00	22101.00	23274.00	22725.00	23591.00
Mean FE	15835.93	18899.03	20494.56	22720.95	24348.72	26015.66	26831.92	28345.96	28983.49	29867.43	30043.99
Max FE	34132.00	29946.00	35685.00	39572.00	36275.00	48905.00	40522.00	45340.00	43551.00	44468.00	41505.00
95 % confidence	882.57	963.27	924.66	987.39	835.16	1026.15	750.56	980.87	766.63	945.23	634.06
Min HE	13554.00	15289.00	17862.00	20967.00	22848.00	25315.00	26185.00	29572.00	31381.00	31841.00	34670.00
Mean HE	19645.80	23519.22	25946.24	28682.01	31199.07	34029.35	35636.52	38600.34	40171.84	43179.94	43006.91
Max HE	50990.00	41425.00	51822.00	63945.00	56522.00	84655.00	53510.00	71318.00	62205.00	93556.00	59979.00
95 % confidence	1508.04	1520.77	1601.16	1828.42	1537.08	2044.16	1444.30	1788.73	1346.45	2125.78	1071.64
Min FEN	1.00	1.00	1.00	1.00	1.00	1.00	1.00	1.00	1.00	1.00	1.00
Mean FEN	3.42	3.73	3.24	3.12	3.13	2.85	2.70	2.78	2.51	2.54	2.00
Max FEN	20.00	12.00	15.00	15.00	12.00	18.00	9.00	10.00	10.00	13.00	6.00
95 % confidence	0.84	0.74	0.67	0.65	0.56	0.58	0.43	0.47	0.40	0.48	0.23

In this specific work the genetic operator under investigation is mating. Table contents are as defined for Table B.2

Table B.21 Table of data from Lamarckian GA calculations, using least squares fitting (LSF) for function evaluation (FE) and a multivariate local minimisation routine, to optimise the orientation variables θ, ψ and ϕ for an icosahedral (Ih) structure when compared to a model image

Mutation rate	0.00	0.10	0.20	0.30	0.40	0.50	0.60	0.70	0.80	0.90	1.00
Convergence	86.00	84.00	85.00	94.00	87.00	88.00	92.00	91.00	92.00	94.00	90.00
Min FE	17321.00	19398.00	22885.00	25762.00	28810.00	26620.00	33392.00	32208.00	34050.00	34995.00	36248.00
Mean FE	23140.38	25964.30	28754.34	32511.13	35636.36	37718.93	40628.33	42621.38	43209.55	44563.16	46331.82
Max FE	30539.00	35083.00	41693.00	49449.00	47986.00	59098.00	59721.00	66799.00	59961.00	60754.00	65022.00
95 % confidence	476.82	659.39	868.36	961.62	917.17	1020.33	1042.37	1162.46	1122.58	908.73	1217.04
Min HE	22988.00	27897.00	31206.00	32819.00	40251.00	41365.00	47018.00	49414.00	58290.00	63171.00	67976.00
Mean HE	31632.37	36284.89	39938.18	46208.60	51705.93	56789.15	62534.24	67814.63	72994.04	76271.79	82697.98
Max HE	47712.00	55329.00	68392.00	90233.00	100892.00	106979.00	106518.00	122930.00	119707.00	103730.00	130918.00
95 % confidence	973.40	1220.89	1526.19	2023.40	2069.73	2167.72	2048.43	2502.45	2346.22	1705.40	2493.54
Min FEN	1.00	1.00	1.00	1.00	1.00	1.00	1.00	1.00	1.00	1.00	1.00
Mean FEN	2.42	2.81	2.49	2.60	2.51	2.27	2.36	2.34	2.13	1.93	2.03
Max FEN	7.00	9.00	9.00	11.00	9.00	10.00	10.00	9.00	7.00	4.00	7.00
95 % confidence	0.32	0.40	0.41	0.40	0.35	0.31	0.32	0.30	0.24	0.15	0.25

In this specific work the genetic operator under investigation is mutation of parents. Table contents are as defined for Table B.3

Table B.22 Table of data from Lamarckian GA calculations, using least squares fitting (LSF) for function evaluation (FE) and a multivariate local minimisation routine, to optimise the orientation variables θ, ψ and ϕ for an icosahedral (Ih) structure when compared to a model image

Mutation rate	0.00	0.10	0.20	0.30	0.40	0.50	0.60	0.70	0.80	0.90	1.00
Convergence	73.00	87.00	87.00	84.00	92.00	91.00	95.00	93.00	95.00	94.00	90.00
Min FE	18891.00	20361.00	24095.00	27445.00	28212.00	30709.00	30461.00	33641.00	37146.00	41339.00	38903.00
Mean FE	24736.36	27893.91	30801.79	33867.45	36686.55	39427.82	43303.39	45598.27	49656.29	51312.86	54574.18
Max FE	33121.00	41023.00	43508.00	55906.00	51956.00	55128.00	67404.00	67557.00	77654.00	78421.00	81567.00
95 % confidence	708.53	739.30	852.94	1013.53	918.81	930.20	1380.13	1216.78	1392.77	1374.61	1378.02
Min HE	24224.00	27951.00	30805.00	34935.00	35507.00	38768.00	43293.00	45805.00	48268.00	50054.00	56070.00
Mean HE	32886.25	37008.46	39616.60	44044.92	47586.64	50321.70	56580.33	59290.28	65089.98	67574.03	72143.01
Max HE	52719.00	67211.00	73177.00	104870.00	87051.00	92793.00	108661.00	88260.00	126520.00	134027.00	141258.00
95 % confidence	1561.57	1628.31	1590.15	2124.84	1826.02	1909.91	2356.82	1672.24	2783.66	2757.40	2732.13
Min FEN	1.00	1.00	1.00	1.00	1.00	1.00	1.00	1.00	1.00	1.00	1.00
Mean FEN	2.68	2.78	2.36	2.39	2.38	2.10	2.39	2.15	2.38	2.14	2.20
Max FEN	9.00	10.00	10.00	14.00	10.00	8.00	12.00	6.00	12.00	9.00	10.00
95 % confidence	0.47	0.46	0.39	0.43	0.37	0.32	0.38	0.24	0.37	0.31	0.29

In this specific work the genetic operator under investigation is mutation of offspring. Table contents are as defined for Table B.4

Table B.23 Table of data from Lamarckian GA calculations, using covariance for function evaluation (FE) and a multivariate local minimisation routine, to optimise the orientation variables θ, ψ and ϕ for an icosahedral (Ih) structure when compared to a model image

Offspring rate	0.00	0.10	0.20	0.30	0.40	0.50	0.60	0.70	0.80	0.90	1.00
Convergence	69.00	75.00	87.00	76.00	78.00	80.00	82.00	84.00	82.00	85.00	90.00
Min FE	11432.00	13567.00	15887.00	17184.00	17306.00	18672.00	21242.00	22297.00	21291.00	21572.00	25167.00
Mean FE	15605.61	18894.67	20585.01	22612.21	23796.10	24718.53	27140.88	28233.13	29293.29	31777.93	32352.54
Max FE	29996.00	37157.00	34682.00	43645.00	38562.00	37459.00	44562.00	40465.00	40835.00	49680.00	56428.00
95 % confidence	880.36	1284.34	817.62	1054.64	902.97	791.79	957.86	833.81	742.74	1123.66	983.69
Min HE	13894.00	16347.00	18771.00	21141.00	20543.00	25497.00	25198.00	29215.00	30972.00	32797.00	34603.00
Mean HE	20177.00	24723.84	26086.26	29412.18	30821.31	33048.06	36521.62	38647.46	40459.45	46413.08	46917.12
Max HE	39304.00	57262.00	47803.00	78872.00	53823.00	69830.00	107713.00	65162.00	64679.00	79878.00	102143.00
95 % confidence	1441.48	2154.25	1318.42	2013.54	1592.18	1559.98	2418.57	1685.16	1484.88	2198.17	2185.11
Min FEN	1.00	1.00	1.00	1.00	1.00	1.00	1.00	1.00	1.00	1.00	1.00
Mean FEN	3.70	4.35	3.33	3.34	2.76	2.54	2.68	2.54	2.28	2.81	2.29
Max FEN	16.00	23.00	13.00	20.00	10.00	11.00	16.00	9.00	9.00	10.00	15.00
95 % confidence	0.87	1.09	0.58	0.71	0.50	0.44	0.53	0.41	0.32	0.47	0.42

In this specific work the genetic operator under investigation is mating. Table contents are as defined for Table B.2

Table B.24 Table of data from Lamarckian GA calculations, using covariance for function evaluation (FE) and a multivariate local minimisation routine, to optimise the orientation variables θ, ψ and ϕ for an icosahedral (Ih) structure when compared to a model image

Mutation rate	0.00	0.10	0.20	0.30	0.40	0.50	0.60	0.70	0.80	0.90	1.00
Convergence	80.00	79.00	81.00	85.00	88.00	95.00	91.00	96.00	90.00	95.00	93.00
Min FE	17802.00	19023.00	23169.00	25505.00	29509.00	31002.00	33385.00	33969.00	37343.00	39543.00	38092.00
Mean FE	22897.69	27278.75	29518.12	33767.79	37166.81	38757.31	42222.18	43644.23	45564.63	47759.52	51508.75
Max FE	31102.00	40649.00	48569.00	59431.00	85820.00	64830.00	74452.00	82050.00	68930.00	74162.00	97863.00
95 % confidence	560.84	939.36	947.14	1321.20	1398.39	1362.72	1397.32	1397.68	1052.90	1349.38	2023.95
Min HE	24412.00	27708.00	30585.00	36023.00	39416.00	44769.00	49022.00	53771.00	54696.00	60480.00	67723.00
Mean HE	31728.38	38525.29	41678.35	48633.19	53895.23	57058.47	63870.69	67498.17	70943.20	77622.12	86111.18
Max HE	57260.00	78937.00	91831.00	118614.00	175351.00	157159.00	126608.00	153147.00	123175.00	144826.00	174673.00
95 % confidence	1219.97	2192.23	2109.70	3152.32	3211.41	2963.31	2822.87	2745.10	2237.12	2550.55	3960.77
Min FEN	1.00	1.00	1.00	1.00	1.00	1.00	1.00	1.00	1.00	1.00	1.00
Mean FEN	2.35	3.04	2.47	2.78	2.68	2.25	2.51	2.24	2.01	2.09	2.46
Max FEN	10.00	11.00	12.00	17.00	23.00	16.00	12.00	14.00	8.00	10.00	12.00
95 % confidence	0.35	0.56	0.44	0.58	0.53	0.43	0.40	0.37	0.24	0.29	0.42

In this specific work the genetic operator under investigation is mutation of parents. Table contents are as defined for Table B.3

Table B.25 Table of data from Lamarckian GA calculations, using covariance for function evaluation (FE) and a multivariate local minimisation routine, to optimise the orientation variables θ, ψ and ϕ for an icosahedral (Ih) structure when compared to a model image

Mutation rate	0.00	0.10	0.20	0.30	0.40	0.50	0.60	0.70	0.80	0.90	1.00
Convergence	74.00	73.00	85.00	80.00	86.00	91.00	86.00	88.00	88.00	94.00	93.00
Min FE	18071.00	19463.00	21575.00	24514.00	26706.00	27681.00	31213.00	35515.00	33089.00	33970.00	40798.00
Mean FE	23911.45	27035.85	29821.91	33535.36	35113.57	38133.26	41575.91	44320.05	48460.65	50549.55	51945.80
Max FE	35763.00	41619.00	40848.00	49098.00	51478.00	64739.00	59133.00	71635.00	79995.00	74724.00	81891.00
95 % confidence	755.18	992.13	831.56	1026.91	1026.87	1057.73	1078.54	1319.57	1596.38	1367.53	1306.48
Min HE	24139.00	27885.00	31788.00	32387.00	37068.00	38773.00	42202.00	45471.00	52746.00	54805.00	57182.00
Mean HE	34070.66	39875.34	41012.32	45719.83	49074.49	52671.78	56891.73	61042.95	68690.51	70151.60	72823.20
Max HE	77086.00	88188.00	74661.00	76998.00	106552.00	100924.00	84795.00	102587.00	143457.00	144110.00	158787.00
95 % confidence	2201.52	2678.67	1836.78	1949.79	2406.53	1948.36	1566.24	2043.52	2890.46	2541.46	2530.62
Min FEN	1.00	1.00	1.00	1.00	1.00	1.00	1.00	1.00	1.00	1.00	1.00
Mean FEN	3.08	3.44	2.52	2.60	2.33	2.27	2.27	2.17	2.56	2.29	2.02
Max FEN	15.00	14.00	12.00	9.00	10.00	9.00	6.00	10.00	14.00	13.00	13.00
95 % confidence	0.66	0.65	0.44	0.39	0.38	0.30	0.24	0.31	0.40	0.33	0.30

In this specific work the genetic operator under investigation is mutation of offspring. Table contents are as defined for Table B.4

B.3.5 Decahedron: LSF, Univariate Local Minimisation

See Tables B.26, B.27 and B.28.

B.3.6 Decahedron: LSF, Powell's Method Local Minimisation

See Tables B.29, B.30 and B.31.

B.3.7 Decahedron: LSF, Multivariate Local Minimisation

See Tables B.32, B.33 and B.34.

B.3.8 Decahedron: Covariance, Multivariate Local Minimisation

See Tables B.35, B.36 and B.37.

Table B.26 Table of data from Lamarckian GA calculations, using least squares fitting (LSF) for function evaluation (FE) and a univariate local minimisation routine, to optimise the orientation variables θ, ψ and ϕ for an decahedral (Dh) structure when compared to a model image

Offspring rate	0.00	0.10	0.20	0.30	0.40	0.50	0.60	0.70	0.80	0.90	1.00
Convergence	39.00	46.00	47.00	60.00	60.00	54.00	59.00	60.00	67.00	72.00	66.00
Min FE	2179.00	2545.00	3170.00	3296.00	3651.00	3626.00	4259.00	4010.00	4288.00	4197.00	4792.00
Mean FE	3465.23	3822.89	4619.89	4965.70	5419.00	5610.63	6112.66	6040.17	6130.78	6441.17	6619.12
Max FE	4909.00	6461.00	7514.00	8239.00	7525.00	7945.00	8727.00	8272.00	9209.00	9068.00	9394.00
95 % confidence	230.37	286.93	275.66	232.14	240.64	311.53	268.54	268.13	262.54	238.06	240.63
Min HE	5927.00	6388.00	7479.00	8032.00	9250.00	9244.00	10045.00	10548.00	11644.00	12090.00	13310.00
Mean HE	10796.38	10579.65	12556.89	13658.70	15242.25	16609.48	18587.31	18006.28	18954.69	20740.88	22042.38
Max HE	16706.00	20920.00	24340.00	27437.00	28151.00	28261.00	31367.00	32280.00	34687.00	38461.00	44240.00
95 % confidence	976.12	1058.89	1031.71	965.35	1072.33	1592.97	1432.72	1335.69	1422.00	1387.60	1447.66
Min FEN	1.00	1.00	1.00	1.00	1.00	1.00	1.00	1.00	1.00	1.00	1.00
Mean FEN	8.54	6.28	7.21	7.10	7.62	8.04	8.98	7.13	7.31	8.07	8.35
Max FEN	19.00	22.00	23.00	25.00	23.00	24.00	23.00	19.00	24.00	22.00	24.00
95 % confidence	1.59	1.54	1.39	1.19	1.28	1.76	1.52	1.30	1.35	1.31	1.24

In this specific work the genetic operator under investigation is mating. Table contents are as defined for Table B.8

Table B.27 Table of data from Lamarckian GA calculations, using least squares fitting (LSF) for function evaluation (FE) and a univariate local minimisation routine, to optimise the orientation variables θ, ψ and ϕ for an decahedral (Dh) structure when compared to a model image

Mutation rate 0.00	0.10	0.20	0.30	0.40	0.50	0.60	0.70	0.80	0.90	1.00	
Convergence	44.00	65.00	67.00	71.00	69.00	80.00	84.00	75.00	85.00	84.00	87.00
Min FE	3371.00	3894.00	4183.00	4863.00	5613.00	5677.00	6368.00	6399.00	6523.00	6905.00	7823.00
Average FE	4432.57	5397.86	6038.81	7147.99	8492.22	8194.70	9150.04	9776.93	10019.02	10585.75	11290.69
Max FE	6292.00	7354.00	8396.00	9743.00	43456.00	11801.00	12296.00	13398.00	13712.00	14483.00	15996.00
95 % confidence	175.06	175.71	230.71	264.34	1047.92	305.86	297.02	368.27	366.43	382.23	388.66
Min HE	9960.00	10948.00	10323.00	12621.00	13750.00	14865.00	15898.00	16989.00	16289.00	17233.00	18520.00
Mean HE	13461.61	17225.00	18220.76	22286.87	24804.22	24154.30	26582.46	28489.07	28837.47	31094.07	32682.52
Max HE	29632.00	30227.00	32311.00	38040.00	60305.00	44708.00	44208.00	48806.00	53188.00	60778.00	55425.00
95 % confidence	1009.90	1059.70	1189.01	1468.42	1849.92	1394.47	1459.48	1644.92	1732.89	1768.85	1713.63
Min FEN	1.00	1.00	1.00	1.00	1.00	1.00	1.00	1.00	1.00	1.00	2.00
Mean FEN	4.09	6.77	6.57	8.58	9.13	7.51	8.04	8.01	7.52	7.68	7.89
Max FEN	21.00	18.00	17.00	20.00	20.00	19.00	18.00	19.00	18.00	19.00	18.00
95 % confidence	1.13	1.09	1.10	1.21	1.16	0.96	0.97	1.00	0.97	0.90	0.81

In this specific work the genetic operator under investigation is mutation of parents. Table contents are as defined for Table B.9

Table B.28 Table of data from Lamarckian GA calculations, using least squares fitting (LSF) for function evaluation (FE) and a univariate local minimisation routine, to optimise the orientation variables θ, ψ and ϕ for a decahedral (Dh) structure when compared to a model image

Mutation rate	0.00	0.10	0.20	0.30	0.40	0.50	0.60	0.70	0.80	0.90	1.00
Convergence	42.00	56.00	53.00	62.00	63.00	56.00	61.00	67.00	70.00	59.00	71.00
Min FE	3773.00	3706.00	4502.00	5042.00	5169.00	5973.00	6135.00	7505.00	7238.00	7760.00	7204.00
Mean FE	4735.67	5537.52	6376.32	7216.92	7886.44	8539.57	9332.36	10469.00	10326.11	11217.14	11869.34
Max FE	5816.00	7194.00	9453.00	10241.00	12148.00	12024.00	13889.00	15452.00	14452.00	16220.00	16615.00
95 % confidence	165.65	203.93	313.93	311.25	381.59	416.99	474.39	463.83	420.48	564.66	521.32
Min HE	10013.00	10428.00	11771.00	13489.00	12525.00	14855.00	13947.00	18066.00	18968.00	19740.00	20234.00
Mean HE	15121.83	17741.84	19518.77	21568.95	22421.03	24390.20	25813.48	30188.84	28252.26	30592.64	32364.77
Max HE	27827.00	33458.00	38430.00	36814.00	39872.00	44223.00	46824.00	52720.00	47501.00	51548.00	51675.00
95 % confidence	1209.65	1304.95	1641.66	1556.47	1659.48	1850.91	1852.06	1994.42	1531.97	2188.45	1930.28
Min FEN	1.00	1.00	1.00	1.00	1.00	1.00	1.00	1.00	1.00	1.00	1.00
Mean FEN	6.14	7.25	7.43	7.40	6.95	7.02	7.00	8.39	6.76	6.53	6.80
Max FEN	23.00	25.00	22.00	19.00	19.00	22.00	18.00	23.00	16.00	17.00	15.00
95 % confidence	1.48	1.39	1.53	1.32	1.24	1.35	1.13	1.24	0.92	1.17	0.98

In this specific work the genetic operator under investigation is mutation of offspring. Table contents are as defined for Table B.10

Table B.29 Table of data from Lamarckian GA calculations, using least squares fitting (LSF) for function evaluation (FE) and a Powell's local minimisation routine, to optimise the orientation variables θ, ψ and ϕ for an decahedral (Dh) structure when compared to a model image

Offspring rate	0.00	0.10	0.20	0.30	0.40	0.50	0.60	0.70	0.80	0.90	1.00
Convergence	51.00	59.00	69.00	65.00	65.00	65.00	74.00	68.00	80.00	63.00	71.00
Min FE	2456.00	2938.00	3294.00	3687.00	3562.00	4438.00	3733.00	4230.00	4099.00	4817.00	4859.00
Mean FE	3560.20	4401.12	4693.57	4996.42	5373.66	5869.88	5966.00	6332.87	6362.96	6649.98	6859.23
Max FE	5265.00	6652.00	8042.00	8376.00	8923.00	8513.00	9394.00	8781.00	8936.00	10670.00	9156.00
95 % confidence	193.94	217.26	251.54	242.77	281.32	238.68	244.02	274.37	205.94	256.53	246.70
Min HE	7597.00	8014.00	9138.00	9778.00	10271.00	11829.00	12062.00	12633.00	13902.00	14392.00	15279.00
Mean HE	12207.33	13998.80	14436.35	15232.86	17347.83	18476.18	19986.15	21877.94	22098.49	23549.59	24732.38
Max HE	20781.00	22603.00	33466.00	28716.00	39908.00	35770.00	39829.00	42116.00	39124.00	50029.00	43470.00
95 % confidence	954.70	955.27	1204.32	1148.86	1619.39	1478.26	1436.73	1724.84	1179.58	1842.34	1602.15
Min FEN	1.00	1.00	1.00	1.00	1.00	1.00	1.00	1.00	1.00	1.00	1.00
Mean FEN	6.88	7.29	6.17	5.63	6.46	6.11	6.39	7.18	6.56	6.67	6.72
Max FEN	19.00	19.00	28.00	18.00	26.00	21.00	25.00	25.00	19.00	27.00	19.00
95 % confidence	1.21	1.11	1.31	1.11	1.47	1.27	1.19	1.39	0.99	1.33	1.12

In this specific work the genetic operator under investigation is mating. Table contents are as defined for Table B.8

Table B.30 Table of data from Lamarckian GA calculations, using least squares fitting (LSF) for function evaluation (FE) and a Powell's local minimisation routine, to optimise the orientation variables θ, ψ and ϕ for an decahedral (Dh) structure when compared to a model image

Mutation rate 0.00	0.10	0.20	0.30	0.40	0.50	0.60	0.70	0.80	0.90	1.00	
Convergence	47.00	65.00	75.00	70.00	82.00	78.00	88.00	87.00	88.00	87.00	90.00
Min FE	3712.00	4086.00	3992.00	4994.00	6000.00	6279.00	6426.00	6362.00	6879.00	8086.00	7712.00
Mean FE	4542.45	5417.11	6373.39	7110.20	7926.37	8624.78	9178.99	9872.30	10440.49	10773.92	11234.43
Max FE	6302.00	7339.00	9412.00	10744.00	11425.00	12438.00	13395.00	15594.00	14722.00	14410.00	17912.00
95 % confidence	166.83	189.17	228.74	275.06	290.76	292.32	339.70	366.98	365.91	337.25	381.55
Min HE	10055.00	12803.00	14314.00	14866.00	15768.00	17822.00	17576.00	18430.00	21504.00	21102.00	21527.00
Mean HE	14731.40	18975.66	22057.49	24121.43	25999.90	27641.87	29082.43	31102.74	32868.09	33010.46	34874.63
Max HE	31413.00	34053.00	39549.00	41021.00	53822.00	50724.00	50348.00	51479.00	52442.00	49972.00	66478.00
95 % confidence	1105.46	1292.25	1320.57	1520.70	1664.00	1593.55	1633.57	1638.51	1607.31	1469.15	1816.84
Min FEN	1.00	1.00	1.00	1.00	1.00	1.00	1.00	1.00	1.00	1.00	1.00
Mean FEN	5.49	5.28	6.45	6.73	6.67	6.81	6.52	6.67	6.85	6.21	6.34
Max FEN	23.00	17.00	20.00	17.00	22.00	20.00	17.00	17.00	16.00	13.00	17.00
95 % confidence	1.25	1.09	1.07	1.09	1.09	0.99	0.90	0.86	0.81	0.70	0.75

In this specific work the genetic operator under investigation is mutation of parents. Table contents are as defined for Table B.9

Table B.31 Table of data from Lamarckian GA calculations, using least squares fitting (LSF) for function evaluation (FE) and a Powell's local minimisation routine, to optimise the orientation variables θ, ψ and ϕ for an decahedral (Dh) structure when compared to a model image

Mutation rate	0.00	0.10	0.20	0.30	0.40	0.50	0.60	0.70	0.80	0.90	1.00
Convergence	53.00	59.00	66.00	68.00	70.00	60.00	77.00	73.00	66.00	78.00	84.00
Min FE	3554.00	4108.00	4770.00	5378.00	5711.00	5497.00	6768.00	7171.00	6679.00	8480.00	8049.00
Mean FE	4675.32	5546.93	6416.50	7301.29	8236.94	8452.77	9683.25	10581.99	11194.12	12034.23	12756.42
Max FE	5670.00	7341.00	8718.00	10917.00	11246.00	12856.00	13477.00	14823.00	16033.00	18808.00	20290.00
95 % confidence	145.11	177.03	243.15	302.72	320.54	434.51	369.89	442.09	502.42	478.09	561.19
Min HE	11790.00	12886.00	13716.00	14995.00	17386.00	17354.00	17230.00	20045.00	22130.00	24184.00	21938.00
Mean HE	17210.92	18103.14	21435.18	23297.51	26312.39	26102.83	29076.52	32048.10	34261.24	37427.42	38050.26
Max HE	30551.00	29342.00	36769.00	46916.00	40747.00	48230.00	44450.00	51500.00	59599.00	70456.00	71312.00
95 % confidence	1166.45	937.38	1530.59	1747.79	1525.75	2048.95	1573.59	1887.84	2292.12	2166.93	2323.26
Min FEN	1.00	1.00	1.00	1.00	1.00	1.00	1.00	1.00	1.00	1.00	1.00
Mean FEN	4.87	4.39	5.77	5.54	6.41	5.22	6.00	6.44	6.56	7.10	6.64
Max FEN	17.00	14.00	18.00	21.00	16.00	19.00	14.00	16.00	17.00	19.00	20.00
95 % confidence	1.10	0.82	1.23	1.21	1.04	1.23	0.86	0.95	1.04	0.94	0.95

In this specific work the genetic operator under investigation is mutation of offspring. Table contents are as defined for Table B.10

Table B.32 Table of data from Lamarckian GA calculations, using least squares fitting (LSF) for function evaluation (FE) and a multivariate local minimisation routine, to optimise the orientation variables θ, ψ and ϕ for a decahedral (Dh) structure when compared to a model image

Offspring rate	0.00	0.10	0.20	0.30	0.40	0.50	0.60	0.70	0.80	0.90	1.00
Convergence	40.00	45.00	56.00	56.00	55.00	61.00	58.00	55.00	64.00	60.00	58.00
Min FE	13876.00	16080.00	17100.00	18603.00	22453.00	23733.00	25295.00	24814.00	25761.00	28046.00	29898.00
Mean FE	21035.18	24569.64	27602.30	31430.96	34032.80	35233.67	36754.66	38886.18	40976.23	41091.55	42979.86
Max FE	29195.00	35959.00	45530.00	53218.00	54899.00	54224.00	53559.00	57576.00	61195.00	66620.00	62073.00
95 % confidence	1402.33	1427.68	1636.75	1958.68	2063.26	1942.95	1802.29	2053.37	2236.26	1682.10	2258.19
Min HE	15066.00	17989.00	17565.00	24990.00	27308.00	29469.00	31317.00	38116.00	37285.00	37201.00	39994.00
Mean HE	27744.65	34122.31	38343.57	45573.36	52718.84	58164.33	63221.14	66418.58	75828.45	76466.75	84028.48
Max HE	44110.00	57579.00	101503.00	109559.00	116758.00	121600.00	140355.00	137133.00	143886.00	200302.00	174798.00
95 % confidence	2588.72	3153.37	3667.35	4486.33	5251.99	5249.65	5835.39	5752.11	6866.15	6368.69	7766.82
Min FEN	1.00	1.00	1.00	1.00	1.00	1.00	1.00	1.00	1.00	1.00	1.00
Mean FEN	7.23	7.40	7.18	7.89	8.38	7.85	7.60	7.80	8.41	7.93	7.93
Max FEN	16.00	18.00	33.00	29.00	27.00	22.00	25.00	22.00	22.00	25.00	24.00
95 % confidence	1.39	1.37	1.40	1.60	1.63	1.42	1.45	1.46	1.46	1.27	1.45

In this specific work the genetic operator under investigation is mating. Table contents are as defined for Table B.2

Table B.33 Table of data from Lamarckian GA calculations, using least squares fitting (LSF) for function evaluation (FE) and a multivariate local minimisation routine, to optimise the orientation variables θ, ψ and ϕ for a decahedral (Dh) structure when compared to a model image

Mutation rate	0.00	0.10	0.20	0.30	0.40	0.50	0.60	0.70	0.80	0.90	1.00
Convergence 46.00	59.00	67.00	71.00	77.00	80.00	73.00	81.00	85.00	84.00	89.00	
Min FE	23147.00	22874.00	24143.00	31436.00	27785.00	34359.00	35870.00	37981.00	37671.00	38491.00	47470.00
Mean FE	30246.39	35317.12	39775.21	45736.30	48039.55	50947.54	55067.05	57714.95	62130.93	61853.32	72590.00
Max FE	41251.00	54292.00	71049.00	69708.00	78722.00	76370.00	76173.00	97134.00	98887.00	97522.00	102113.00
95 % confidence	1146.70	1708.48	2157.45	2088.58	2328.63	2329.92	2460.69	2608.37	2836.17	2515.21	2789.19
Min HE	30447.00	34147.00	38848.00	41081.00	46783.00	50649.00	57185.00	55619.00	61092.00	71139.00	81382.00
Mean HE	50346.76	60423.49	73500.19	82042.18	88617.03	95019.45	102382.34	108025.89	115138.59	115066.13	141158.72
Max HE	76464.00	131286.00	138822.00	143663.00	200287.00	177291.00	204978.00	251085.00	229866.00	211685.00	292111.00
95 % confidence	3475.30	5302.69	6490.08	6004.91	6747.95	6200.70	6825.74	7464.67	7321.00	6731.88	9364.09
Min FEN	1.00	1.00	1.00	1.00	1.00	1.00	1.00	1.00	1.00	1.00	1.00
Mean FEN	5.96	7.32	7.88	9.01	8.26	7.80	7.86	7.46	7.86	6.74	8.67
Max FEN	16.00	28.00	23.00	20.00	27.00	21.00	18.00	23.00	21.00	17.00	19.00
95 % confidence	1.20	1.48	1.43	1.25	1.25	1.07	1.04	0.99	1.00	0.79	0.93

In this specific work the genetic operator under investigation is mutation of parents. Table contents are as defined for Table B.3

Table B.34 Table of data from Lamarckian GA calculations, using least squares fitting (LSF) for function evaluation (FE) and a multivariate local minimisation routine, to optimise the orientation variables θ, ψ and ϕ for a decahedral (Dh) structure when compared to a model image

Mutation rate	0.00	0.10	0.20	0.30	0.40	0.50	0.60	0.70	0.80	0.90	1.00
Convergence	35.00	50.00	47.00	59.00	57.00	60.00	60.00	58.00	64.00	76.00	68.00
Min FE	22934.00	24829.00	26323.00	31810.00	30758.00	32950.00	29032.00	37711.00	40218.00	42237.00	44689.00
Mean FE	31755.69	36011.90	39554.43	44462.15	49068.00	52980.58	57627.22	63311.28	66258.31	71235.68	68709.28
Max FE	42300.00	52699.00	64367.00	78072.00	82585.00	83212.00	94115.00	94436.00	121566.00	138668.00	111390.00
95 % confidence	1655.47	2217.26	2176.09	2156.56	3057.12	2802.98	3307.30	3524.48	4386.42	4141.27	3534.89
Min HE	27493.00	32171.00	37316.00	41009.00	43924.00	50926.00	53565.00	59625.00	59695.00	71615.00	66726.00
Mean HE	50374.46	62455.96	70561.38	77345.69	83976.42	92943.23	100487.98	111928.05	114931.72	127413.61	124313.75
Max HE	83033.00	151340.00	161785.00	163028.00	161603.00	184403.00	161518.00	192866.00	241905.00	236634.00	210377.00
95 % confidence	4234.76	6319.31	6429.13	6275.57	6697.88	6689.00	7183.95	8772.44	9731.82	9335.90	8636.56
Min FEN	1.00	1.00	1.00	1.00	1.00	1.00	1.00	1.00	1.00	1.00	1.00
Mean FEN	6.09	7.38	7.45	6.90	6.98	7.23	7.27	7.97	7.38	7.68	6.72
Max FEN	15.00	23.00	28.00	20.00	21.00	22.00	18.00	20.00	24.00	21.00	18.00
95 % confidence	1.44	1.67	1.64	1.24	1.34	1.24	1.18	1.30	1.31	1.20	1.09

In this specific work the genetic operator under investigation is mutation of offspring. Table contents are as defined for Table B.4

Table B.35 Table of data from Lamarckian GA calculations, using covariance for function evaluation (FE) and a multivariate local minimisation routine, to optimise the orientation variables θ, ψ and ϕ for a decahedral (Dh) structure when compared to a model image

Offspring rate	0.00	0.10	0.20	0.30	0.40	0.50	0.60	0.70	0.80	0.90	1.00
Convergence	1.00	7.00	9.00	12.00	12.00	9.00	11.00	16.00	9.00	17.00	20.00
Min FE	22323.00	20098.00	17143.00	19546.00	20694.00	26965.00	24157.00	24529.00	31812.00	30221.00	28167.00
Mean FE	22323.00	25078.86	22811.00	30866.00	30622.25	34855.33	33764.45	41068.00	39425.67	39218.76	41299.20
Max FE	22323.00	30459.00	35215.00	38093.00	37570.00	46102.00	50396.00	62554.00	51008.00	48708.00	57116.00
95 % confidence	–	2956.85	4015.87	3455.74	3093.67	4458.19	4294.52	4590.32	4072.97	2427.11	3414.91
Min HE	33342.00	23741.00	19131.00	24128.00	28812.00	35189.00	36755.00	37065.00	46560.00	47341.00	44232.00
Mean HE	33342.00	35658.43	28445.00	46315.42	46623.50	55723.33	54207.45	75289.25	78978.33	75499.94	76674.75
Max HE	33342.00	52530.00	46019.00	62778.00	63935.00	83876.00	102289.00	121016.00	153895.00	134814.00	165679.00
95 % confidence	–	7542.16	6133.64	6851.91	8445.01	11485.38	10247.69	13143.10	20484.56	10792.13	11875.64
Min FEN	9.00	4.00	1.00	1.00	1.00	3.00	2.00	2.00	1.00	2.00	1.00
Mean FEN	9.00	8.57	3.89	7.67	6.75	8.00	5.18	8.69	7.22	7.18	6.85
Max FEN	9.00	14.00	11.00	13.00	14.00	18.00	14.00	18.00	18.00	21.00	19.00
95 % confidence	–	2.77	2.60	2.16	2.43	3.20	1.94	2.33	3.23	2.29	2.21

In this specific work the genetic operator under investigation is mating. Table contents are as defined for Table B.2

Table B.36 Table of data from Lamarckian GA calculations, using covariance for function evaluation (FE) and a multivariate local minimisation routine, to optimise the orientation variables θ, ψ and ϕ for a decahedral (Dh) structure when compared to a model image

Mutation rate	0.00	0.10	0.20	0.30	0.40	0.50	0.60	0.70	0.80	0.90	1.00
Convergence	8.00	8.00	12.00	19.00	14.00	25.00	24.00	30.00	27.00	29.00	34.00
Min FE	23456.00	25026.00	24882.00	28834.00	28463.00	31287.00	29046.00	40244.00	41987.00	44599.00	47621.00
Mean FE	30107.63	34049.63	35801.50	41933.42	48766.57	48765.92	56597.58	55665.80	60889.93	67002.93	67586.56
Max FE	39729.00	43642.00	42340.00	59618.00	72544.00	62951.00	83240.00	76144.00	87577.00	98953.00	90137.00
95 % confidence	5130.15	4298.82	3000.34	3896.39	6242.55	3660.62	5467.20	3254.21	3937.12	4459.43	3732.23
Min HE	30509.00	33294.00	35058.00	46526.00	45727.00	48300.00	59119.00	66981.00	76460.00	72041.00	71530.00
Mean HE	44698.50	50998.75	62115.75	75190.53	97039.43	92796.88	115705.17	103617.73	114973.63	126648.97	127111.12
Max HE	66021.00	69974.00	98731.00	140918.00	173584.00	137547.00	202759.00	160934.00	193482.00	214193.00	206369.00
95 % confidence	8762.50	8654.53	9839.64	12364.43	18846.40	11105.34	16711.74	9690.03	11357.57	12722.12	11148.29
Min FEN	1.00	1.00	1.00	2.00	1.00	1.00	1.00	2.00	2.00	2.00	2.00
Mean FEN	4.50	4.75	4.92	6.53	9.14	6.88	9.38	6.37	7.07	7.59	7.53
Max FEN	12.00	10.00	9.00	19.00	19.00	16.00	20.00	14.00	17.00	15.00	16.00
95 % confidence	2.59	2.18	1.33	2.33	2.78	1.72	2.48	1.23	1.55	1.34	1.30

In this specific work the genetic operator under investigation is mutation of parents. Table contents are as defined for Table B.3

Table B.37 Table of data from Lamarckian GA calculations, using covariance for function evaluation (FE) and a multivariate local minimisation routine, to optimise the orientation variables θ, ψ and ϕ for a decahedral (Dh) structure when compared to a model image

Mutation rate	0.00	0.10	0.20	0.30	0.40	0.50	0.60	0.70	0.80	0.90	1.00
Convergence	13.00	13.00	10.00	11.00	16.00	14.00	13.00	13.00	11.00	8.00	9.00
Min FE	20111.00	24244.00	33618.00	30398.00	31471.00	38879.00	41054.00	39118.00	41044.00	40288.00	49522.00
Mean FE	29586.31	32344.23	39884.40	44191.27	45461.81	51076.79	54800.08	55255.08	54554.09	54430.88	68801.67
Max FE	37534.00	50626.00	47636.00	55681.00	61694.00	68938.00	74137.00	92879.00	80355.00	74416.00	92113.00
95 % confidence	2849.27	3507.86	3217.49	5490.57	4141.59	5467.06	7159.77	7610.21	7269.28	7894.97	9826.62
Min HE	30365.00	38459.00	53406.00	42932.00	49714.00	58441.00	59201.00	68155.00	68082.00	73301.00	76372.00
Mean HE	46611.85	51918.38	62706.90	76188.36	76066.75	83803.71	91488.23	94774.54	94198.73	103953.25	118658.00
Max HE	63895.00	90386.00	79072.00	109332.00	115732.00	112880.00	150558.00	147955.00	130035.00	125347.00	137653.00
95 % confidence	6663.73	7235.91	5214.90	12612.06	8035.05	10243.63	15836.11	13822.58	14421.43	13189.54	12588.97
Min FEN	1.00	1.00	2.00	1.00	1.00	1.00	1.00	1.00	1.00	1.00	2.00
Mean FEN	4.69	4.46	5.00	6.45	5.44	5.00	4.85	4.38	3.55	3.38	4.78
Max FEN	9.00	13.00	8.00	12.00	14.00	10.00	12.00	12.00	8.00	7.00	7.00
95 % confidence	1.53	1.90	1.24	2.37	1.63	1.58	2.00	1.84	1.66	1.57	1.12

In this specific work the genetic operator under investigation is mutation of offspring. Table contents are as defined for Table B.4

Appendix C
First-Principles Calculations of an Au_{16} Nanocluster on Graphite

C.1 Γ-Point Versus 13 k-Point: A Comparison

See Table C.1.

Table C.1 Comparison of E_{tot} for Γ-point and 13 **k**-point system calculations, with C atoms unconstrained within 3.5 Å of the Au_{16} cluster

	GR2	GR2-h	GR2-2h	GR2-3h
ΔE_{tot} (eV)				
Au_{16} (0001) face	2.79 (0.21 %)	2.25 (0.17 %)	1.96 (0.15 %)	2.20 (0.17 %)
Au_{16} (111) face	2.78 (0.21 %)	2.22 (0.17 %)	1.91 (0.14 %)	2.20 (0.17 %)

The value ΔE_{tot} is calculated by subtraction of the total energy of the latter (13 **k**-point) from the former (Γ-point). Percentage of overall value is also given in brackets

C.2 Comparison of the PBE, vdW-DF and vdW-DF2 Exchange-Correlation Functionals

See Table C.2.

C.3 Bending of the Graphite Surface

See Table C.3.

A. J. Logsdail, *Computational Characterisation of Gold Nanocluster Structures*, Springer Theses, DOI: 10.1007/978-3-319-01493-7,

Table C.2 Au_{16} cluster adsorbed on (111) face on HOPG with (GR2-2h and without (GR2) multiple surface defects using the vdW-DF and vdW-DF2 exchange-correlation functionals compared to PBE functional

		GR2	GR2-2h
PBE			
E_a	(eV)	0.80	3.79
Au-X	(Å)	3.33	1.99
$d_\perp$(Au)	(Å)	3.38	0.78
$d_\perp$(X)	(Å)	0.07	0.58
vdW-DF			
E_a		1.35	3.10
Au-X		3.32	2.00
$d_\perp$(Au)		3.36	0.83
$d_\perp$(X)		0.05	0.59
vdW-DF2			
E_a		1.47	2.91
Au-X		3.24	2.01
$d_\perp$(Au)		3.27	0.84
$d_\perp$(X)		0.04	0.58

C atoms are unconstrained within 3.5 Å of the Au_{16} cluster. In all cases X is considered to be the nearest neighbouring C atom to the Au_{16} cluster, and all C atoms are unconstrained within 3.5 Å. Adsorption energy, (E_a), vertical displacements from the upper graphene plane [$d_\perp$(Au) and $d_\perp$(X)] and the nearest neighbour (Au-X) distances. For perfect HOPG surface the Au_{16} cluster lies closer than using the PBE functional; however the reverse trend is found when a defect is present

Table C.3 Maximum displacement [$d_\perp$(X)]/Å of C atoms from cluster-free HOPG plane using the PBE, vdW-DF and vdW-DF2 exchange-correlation functionals

	GR2	GR2-2h
PBE	0.068	0.059
vdW-DF	0.063	0.052
vdW-DF2	0.053	0.046

Constraints, identical to those used in the presence of the (111) face-deposited Au_{16} cluster, are used beyond 3.5 Å

C.4 Structural Changes in a Spin-Unpaired Graphite Surface

See Fig. C.1.

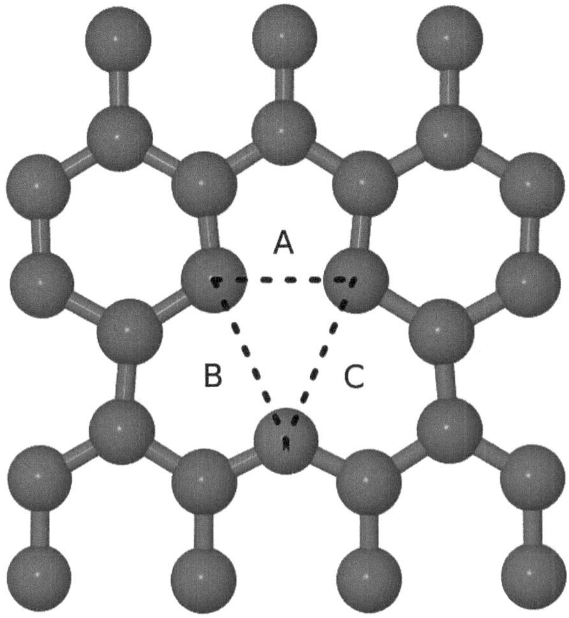

Fig. C.1 Fragment of the deformed HOPG surface with one defect (GR-h), with an initial magnetic moment, illustrated here for a 7.0 Å radius of unconstrained atoms. Distance A (2.12 Å) is significantly less than distances B and C (2.59 and 2.57 Å, respectively); for the spin-paired system the distances A, B and C are 2.38, 2.61 and 2.55 Å, respectively. For GR-h with constraints beyond 3.5 Å it was found that the HOPG plane could not distort as necessary to support the spin-unpaired system. For spin-paired. the distances A, B and C were found to be 2.45, 2.54 and 2.59 Å, respectively. Reprinted with permission from Ref. [1]. Copyright 2011 American Chemical Society

C.5 Results for Calculations with Carbon Atoms Constrained Beyond 7 Å from the Au_{16} Cluster

See Tables C.4, C.5 and C.6.

Table C.4 Au_{16} cluster on HOPG with and without multiple surface defects

		GR2	GR2-h	GR2-2h	GR2-3h
Au (111) face					
E_a	(eV)	0.71	3.67	3.64	5.04
Au-X	(Å)	3.02	2.08	2.02	2.00
$d_\perp$(Au)	(Å)	3.38	2.10	1.05	0.92
$d_\perp$(X)	(Å)	0.36	0.82	0.92	0.97
Δq	(e)	0.68	0.45	−0.13	−0.10
Au (0001) face					
E_a		0.68	3.40	3.56	4.96
Au-X		3.68	2.09	2.01	1.99
$d_\perp$(Au)		3.78	2.05	1.03	1.03
$d_\perp$(X)		0.27	0.69	0.90	0.89
Δq		0.68	0.40	−0.13	−0.08

C atoms are unconstrained within 7.0 Å of the Au_{16} cluster. Measurements are as defined for Table C.2, and the charge transfer (Δq) is also given. $\Delta q > 0$ implies charge transfer to the Au cluster. When compared to our previous observations for constraints applied beyond 3.5 Å from the Au_{16} cluster (Chap. 4.4), the order of results for Au_{16} (111) adsorption compared to (0001) is different. With increased HOPG freedom (7.0 Å) E_a is greater for Au_{16} (111) face adsorption (5.04 compared to 4.96 eV for adsorption on the Au_{16} (0001) face), matching observations for an unconstrained distance of 3.5 Å. $d_\perp$(X) is considerably larger for Au_{16} (111) face adsorption than (0001) for all substrate arrangements with the greater degree of freedom (7.0 Å) in the HOPG layers; the opposite is seen for 3.5 Å

Table C.5 Structural analysis of Au_{16} cluster on HOPG graphite with and without multiple surface defects: as defined

		GR2	GR2-h	GR2-2h	GR2-3h
Au (111) face					
N_{Au-Au}		42	42	41	42
$\bar{r}_{Au-Au}$	(Å)	2.80	2.80	2.82	2.84
Symmetry		T_d	C_1	C_1	C_1
ΔE	(eV)	0.01	0.13	−0.37	−0.30
Au (0001) face					
N_{Au-Au}		41	42	39	42
$\bar{r}_{Au-Au}$	(Å)	2.79	2.80	2.82	2.84
Symmetry		T_d	C_s	C_1	C_1
ΔE	(eV)	0.47	4-0.03	−0.89	−0.46

C atoms are unconstrained within 7.0 Å of the Au_{16} cluster. The number of Au-Au bonds below 3.5 Å in length (N_{Au-Au}), average (mean) Au-Au bond length ($\bar{r}_{Au-Au}$) for all bonds in N_{Au-Au}, cluster symmetry and the difference in energy between the isolated Au_{16} structure (T_d symmetry) and the energy of the isolated Au_{16} cluster as adsorbed (ΔE) are given, where $\Delta E > 0$ implies the adsorbed structure is more stable

Table C.6 Interlayer distances and displacement from the HOPG plane of the top C atom ($d_{\perp}$(X)) for isolated (i.e. cluster free) perfect graphite (GR2), one surface vacancy (GR2-h), two surface vacancies (GR-2h) and three surface vacancies (GR2-3h)

		GR2	GR2-h	GR2-2h	GR2-3h
Constraints beyond 3.5 Å radius					
Au (111) face					
Interlayer separation	(Å)	3.360	3.359	3.358	3.357
$d_{\perp}$(X)	(Å)	0.068	0.063	0.059	0.055
Au (0001) face					
Interlayer separation		3.366	3.366	3.365	3.363
$d_{\perp}$(X)		0.075	0.071	0.069	0.066
Constraints beyond 7.0 Å radius					
Au (111) face					
Interlayer separation		3.539	3.523	3.505	3.555
$d_{\perp}$(X)		0.320	0.299	0.197	0.275
Au (0001) face					
Interlayer separation		3.613	3.601	3.503	3.520
$d_{\perp}$(X)		0.269	0.285	0.186	0.277

Constraints of both 3.5 and 7.0 Å are enforced to match the calculations in the presence of soft-landed Au_{16} clusters. Before geometry optimisation these values equal to 3.35 and 0 Å, respectively

Reference

1. A.J. Logsdail, R.L. Johnston, Predicting the optical properties of core-shell and janus segregated au-m nanoparticles (m = ag, pd). J. Phys. Chem. C **116**(44), 23616–23628 (2012)

Appendix D
Extinction Spectra Plots from Optical Calculations

D.1 Effect of Varying the Incoming Radiation Angle on the Extinction Spectra for a Core-Displaced $Pd_{core}Au_{shell}$ Particle

See Fig. D.1.

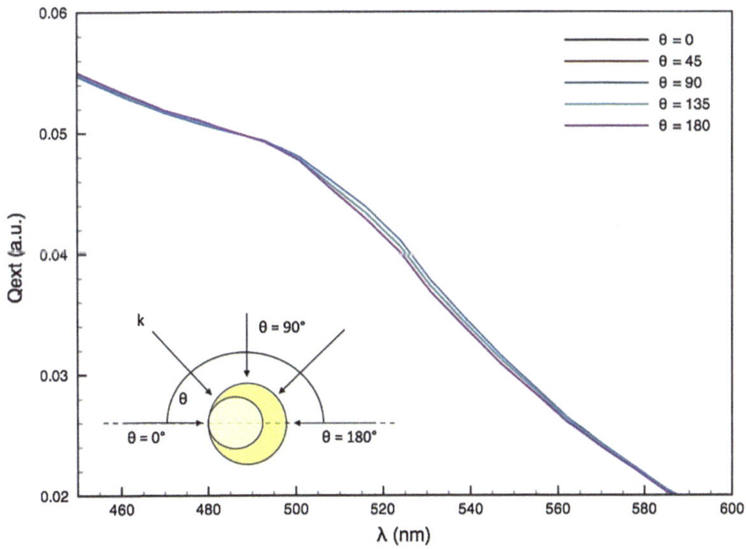

Fig. D.1 Q_{ext} plotted for a $Pd_{core}Au_{shell}$ particle, with the core-displaced, at varying angles of incoming radiation, θ. A key is given; refractive indices have been interpolated from the literature. Reprinted with permission from Ref. [1]. Copyright 2012 American Chemical Society

A. J. Logsdail, *Computational Characterisation of Gold Nanocluster Structures*, Springer Theses, DOI: 10.1007/978-3-319-01493-7,

D.2 Enlarged Plot of the Extinction Spectra for Core-Shell $(AgAu)_N$ Particles

See Fig. D.2.

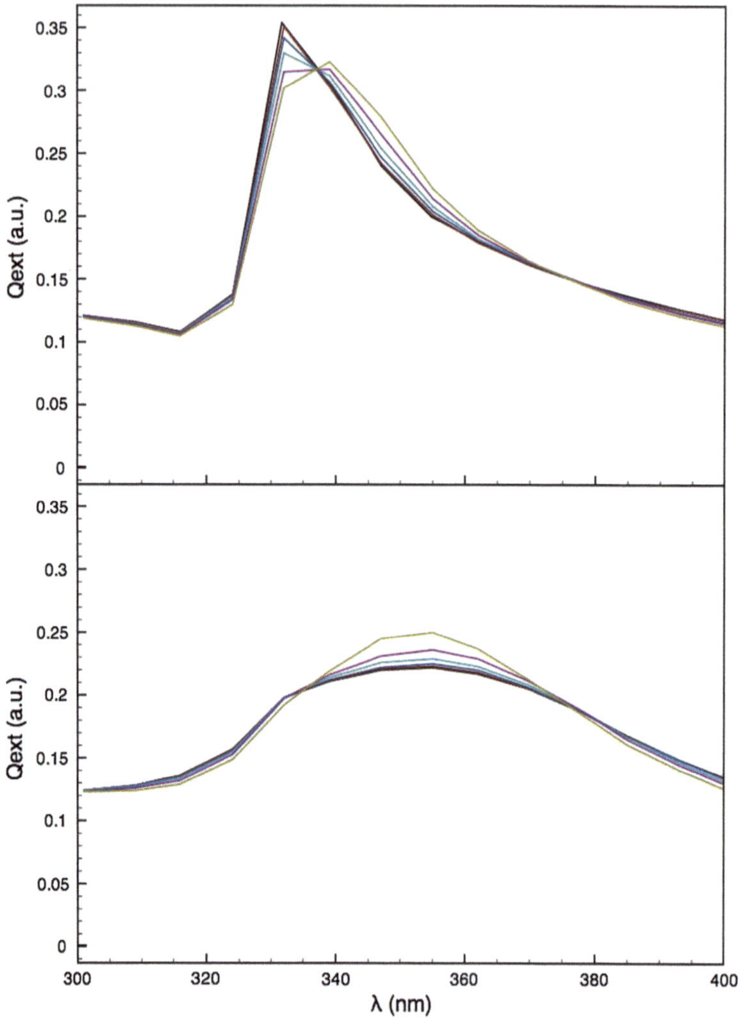

Fig. D.2 Enlarged plot of Fig. 5.18, concentrated on the Ag spectral feature. *Top* Q_{ext} for $Au_{core}Ag_{shell}$ spheres with increasing displacement of the core. *Bottom* Q_{ext} for $Ag_{core}Au_{shell}$ with core displacement. The core-centered particle is plotted in *black*, with displacements of 0.1, 0.2, 0.3, 0.4 and 0.5 nm (i.e. up to touching the surface) plotted in *red*, *blue*, *green*, *purple* and *olive*, respectively. Refractive indices have been interpolated from the literature

Reference

1. A.J. Logsdail, J. Akola, Interaction of Au_{16} nanocluster with defects in supporting graphite: a density-functional study. J. Phys. Chem. C **115**(31), 15240–15250 (2011)

MIX
Papier aus verantwortungsvollen Quellen
Paper from responsible sources
FSC® C105338

If you have any concerns about our products,
you can contact us on
ProductSafety@springernature.com

In case Publisher is established outside the EU,
the EU authorized representative is:
Springer Nature Customer Service Center GmbH
Europaplatz 3, 69115 Heidelberg, Germany

Printed by Libri Plureos GmbH
in Hamburg, Germany